DEVELOPMENT PROJECTS AS POLICY EXPERIMENTS

As international agencies' assistance strategies become more complex, their methods of planning and administration become less effective. Neither the rationalistic techniques of planning and management they adopted during the 1960s and 1970s to control development activities nor the structural adjustment models they used during the 1980s and 1990s to reform economic policies, encouraged the flexibility, experimentation and social learning that are crucial to implementing successfully complex and uncertain development activities. Urgent reorientation of development programmes and continuous testing and verification is required if development activity is to cope effectively with the uncertainty and complexity of the development process.

An *adaptive approach* is needed, an approach which relies on strategic planning, administrative procedures that facilitate innovation, responsiveness and experimentation and on decision-making processes that join learning with action.

Following practical testing and reformulation of ideas presented in the first edition, this up-dated text offers new examples and extended coverage of participatory and strategic management, from individual development projects to economic reform policies such as structural adjustment and privatization.

This book will be of interest to development practitioners and students and academics of the sociology, economics and management of development.

Dennis A. Rondinelli is Professor of International Business at the Kenan-Flagler Business School, and Director of the International Private Enterprise Research Centre at the Kenan Institute of Private Enterprise, University of North Carolina at Chapel Hill.

DEVELOPMENT AND UNDERDEVELOPMENT

Series editors: Ray Bromley and Gavin Kitching

In the same series

DEVELOPMENT AND UNDERDEVELOPMENT IN HISTORICAL PERSPECTIVE:
POPULISM, NATIONALISM AND INDUSTRIALIZATION
Gavin Kitching

DEVELOPMENT AND THE ENVIRONMENTAL CRISES:
RED OR GREEN ALTERNATIVES?
Michael Redclift

REGIONS IN QUESTION:
SPACE, DEVELOPMENT THEORY AND REGIONAL POLICY
Charles Gore

THE FRAGMENTED WORLD:
COMPETING PERSPECTIVES ON TRADE, MONEY AND CRISES
Chris Edwards

TRANSNATIONAL CORPORATIONS AND UNEVEN DEVELOPMENT:
THE INTERNATIONALIZATION OF CAPITAL AND THE THIRD WORLD
Rhys Jenkins

SOCIALISM AND UNDERDEVELOPMENT
Ken Post and Phil Wright

LATIN AMERICAN THEORIES OF DEVELOPMENT AND UNDERDEVELOPMENT
Christóbal Kay

METROPOLIS 2000:
PLANNING, POVERTY AND POLITICS
Thomas Angotti

DEVELOPMENT PROJECTS AS POLICY EXPERIMENTS

An adaptive approach to development administration

Dennis A. Rondinelli

Second edition

London and New York

First published 1993
by Routledge
11 New Fetter Lane, London EC4P 4EE

Simultaneously published in the USA and Canada
by Routledge
29 West 35th Street, New York, NY 10001

Typeset in Garamond by LaserScript, Mitcham, Surrey
Printed and bound in Great Britain by
Biddles Ltd, Guildford and King's Lynn

British Library Cataloguing in Publication Data
A catalogue reference for this book is available from the British Library.

ISBN 0–415–06622–0 Hb
ISBN 0–415–06623–9 Pb

Library of Congress Cataloging in Publication Data

Rondinelli, Dennis A.
Development projects as policy experiments: an adaptive approach to development administration/Dennis A. Rondinelli. – 2nd ed.
p. cm. – (Development and underdevelopment)
Includes bibliographical references and index.
ISBN 0–415–06622–0 — ISBN 0–415–06623–9
1. Economic development projects – Developing countries – Managment. I. Title. II. Series.
HC59.72E44R67 1993
338.9′0068–dc20 92-17773
CIP

CONTENTS

FIGURES

PREFACE

As the development strategies of international assistance agencies and poor countries became more complex over the past few decades, the methods of planning and managing development projects became less effective. Neither the rationalistic techniques of planning and management adopted during the 1960s and 1970s to control development activities nor the structural adjustment models used during the 1980s and 1990s to reform economic policies encouraged the flexibility, experimentation and social learning that are crucial to successfully implementing complex and uncertain development activities.

The conventional planning and administrative methods used by international assistance agencies and governments in developing countries often resulted in costly and ineffective analysis, greater inconsistency and uncertainty, and the delegation of important development activities to technical experts. They have led international agencies and governments to make inappropriate and sometimes harmful interventions because the intended beneficiaries were not encouraged to participate in decision-making and because administrators were discouraged from detecting and correcting errors. Structural adjustment policies evolved from methods of analysis that often ignored or discounted the political, institutional and social aspects of development. Both approaches suffer from the difficulties of defining development objectives concisely, lack of appropriate data, inadequate understanding of local social and cultural conditions, ineffective means of controlling behavior, the dynamics of political interaction, and low levels of administrative capacity in many developing countries.

In this book I argue that development administration must be reoriented to cope more effectively with the inevitable uncertainty and complexity of the development process. One of the most promising ways is to use an *adaptive approach* that relies on strategic planning, on administrative procedures that facilitate innovation, responsiveness and experimentation, and on decision-making processes that join learning with action.

It is appropriate that a book recommending a learning approach to planning and administration is the result of my own learning and experimentation over more than two decades. The ideas presented here build on those found in the first edition. I have reformulated them a number of times in articles and books and tested them in development projects for which I have served as an advisor. As is the case with any ideas that are the result of learning and experimentation, I expect that they will change as I and others continue to test them in the future. Thus, I see them as *propositions* about the nature of development that can provide guides to action, but are still evolving and need continuous testing and verification in new and changing circumstances.

Although I have reformulated my ideas about development administration, my conception of the development process has remained quite consistent with the one I offered more than a decade ago:

> The essence of development is expansion of participation in economic activities through the creation of social and economic systems that draw larger numbers of people into processes of production, exchange, and consumption, that involve greater numbers in entrepreneurship and employment, that increase levels of income for the poorest groups and reduce disparities between rich and poor so that a larger majority of people can obtain basic goods, save and invest and gain access to services necessary to enrich the quality of their lives. Development is a process of expanding the productive capacity of public and private organizations, large and small firms, rural and urban regions of a country at a steady pace. It involves stimulating the use of potentially productive resources, adapting appropriate technologies and institutions to traditional as well as modern communities, transforming subsistence agricultural and rural sectors into employment- and income-generating elements of the national economy, and providing social services and facilities that allow people not only to satisfy basic needs but also to develop their productive capacity and human potential.
>
> (Rondinelli and Ruddle 1978: v)

If my conception has changed it is only in the depth of my conviction that strengthening human resources and the institutions that give individuals freedom to realize their human potential is the foundation of development in both poor and rich countries. Ultimately, public or private institutions can only be assessed by their effectiveness in promoting individual development and social responsibility. Many of the problems with implementing economic assistance programs over the past half century can be attributed, I believe, to a recurring failure to recognize that human development is the primary objective of and foundation for all development activities.

Discussions with many people involved directly in development policy making and implementation helped shape the ideas in this book. The references identify those from whose experience and thinking I have learned the most. For reading and commenting on the first edition, I am especially grateful to Marcus Ingle, David Korten and Edward Rizzo. Ray Bromley read and reacted to manuscripts for both the first and second editions. I retain full responsibility, of course, for the conclusions and interpretations.

Dennis A. Rondinelli
Chapel Hill, North Carolina

1

THE PROBLEM OF DEVELOPMENT ADMINISTRATION

Coping with complexity and uncertainty

Experience with development during the past half century has led to three fundamental discoveries. The first is that conventional theories of development based merely on accelerating economic growth have not achieved their intended goals. Despite progress in raising the levels of per capita gross national product (GNP) in many developing nations, disparities in living conditions between rich and poor countries and between the richest and poorest people within developing countries continue to widen. The trickle-down and spread effects that were expected to follow increases in industrial production during the 1950s and 1960s either did not occur or did little to alleviate widespread poverty. Thus, they did not establish a base for sustained economic growth and human development.

New and more complex strategies emerged during the 1970s to reduce the dependence of poor countries on richer ones, spread the benefits of development to lagging regions within developing countries, and increase the productivity and income of the poorest groups. Despite success in some developing countries, however, the absolute numbers of the poor increased. These strategies were displaced during the late 1980s and early 1990s with structural adjustment policies that sought to reorder national economies into market systems, promote private enterprise and encourage international trade and investment. Although many of the reforms prescribed by structural adjustment policies were needed to create an economic environment conducive to development, it remains to be seen how effective they will be in transforming planned economies into market economies, and in improving the incomes and living conditions of the poor in developing countries.

Second, if the success of development assistance from rich to poor countries is measured by the ability of bilateral and multilateral aid organizations to implement programs and projects with sustainable benefits or by their progress in strengthening the institutional capacity for developing countries to undertake their own development activities, the results have been disappointing. For example, assessments by the US Agency for International Development (USAID) indicate that foreign assistance has had a lackluster record in promoting sustainable economic and social progress in developing countries over the past 40 years. Former USAID administrator Alan Woods noted that of the 95 less-developed countries that his agency assisted, only 19 maintained consistent economic growth from 1950 to 1987 (USAID 1989). A study commissioned by USAID found that project sustainability – the degree to which assisted activities remained active or continued delivering benefits to people after international funding ended – was extremely low. In a sample of 212 projects only 11 percent had a strong probability of being sustained after aid was terminated and 26 percent had poor prospects for providing long-term benefits (Kean *et al.* 1987). A review of sixty-two completed health projects found that "more than half of the projects either had failed before project completion or were unlikely to be sustained following termination of US support" (USAID 1988: vii).

The World Bank has also had a poor record of strengthening those institutions that sustain the benefits of development assistance. An assessment of nearly 1,700 loans for projects with institutional development components carried out between 1978 and 1988 showed that the share with substantial achievements declined from 42 to 35 percent; the share with partial achievements declined from 48 to 44 percent (Paul 1989).

A third fundamental discovery is that as strategies for economic and social development became more complex, the success of development policies, programs and projects became less certain. Rationalistic planning and management methods adopted by development assistance organizations during the 1960s and 1970s were of limited use in coping with the uncertainty and complexity of development problems or in responding to the needs of people in developing countries. And there is little evidence that those adopted during the 1980s and 1990s were or will be substantially better.

Experience during the past half century indicates that rapidly changing and disparate theories of economic development have been, and will continue to be, uncertain propositions that are shaped by complex processes of political interaction and social learning. But the planning and administrative procedures used by international organizations and governments of developing countries to implement development projects have never adequately reflected these underlying uncertainties. Nor have those who have applied them recognized explicitly that all development policies are really social experiments. Governments and international organizations still attempt to use rationalistic planning and management techniques to control development activities in order to achieve their own ends, rather than encouraging the flexibility, experimentation, and social learning that are essential to helping the intended beneficiaries achieve *their* objectives.

This book explores the divergence between the nature of the development process and the practice of development administration. It examines a major dilemma of development administration: planners and managers, working in bureaucracies that seek to control rather than to facilitate development, must cope with increasing uncertainty and com- plexity; but their methods and procedures inhibit the kinds of analysis and planning that are most appropriate for dealing with development problems effectively.

THE GAP BETWEEN THEORY AND REALITY

Although the rhetoric of development policy has changed drastically over the past half century, perspectives on the nature of development planning and administration have changed very little. In the 1950s and 1960s, development planners prescribed long-range, comprehensive, national planning and centrally controlled, "top-down" systems of decision-making to formulate and implement development policies. During the 1970s most international development organizations and governments in developing countries adopted what Lindblom (1965) calls a "synoptic" approach to decision-making. As he points out, those who held a *rationalistic* view of decision-making believed that complex social problems could be understood through systematic analysis and solved through comprehensive planning. They assumed the existence of authoritative and objective decision-makers whose actions could, if they were carried out correctly, solve economic and social problems. They believed that exhaustive analysis would lead to a concise definition of problems and generate alternatives from which optimal and correct policy choices could be made. They further believed that they could construct models or theories of social change to aid in problem definition and policy formulation, that the resulting policies would respond adequately to human needs, and that there was a direct relationship between government action and the solution of social problems (Braybrooke and Lindblom 1970).

Underlying these methods was the assumption that plans emanating from rational analysis had to be carried out through a hierarchical structure of authority and a comprehensive system of rules and regulations. Deviations would be detrimental to achieving development objectives. Conflicts over goals, values or courses of action were seen, therefore, as adverse and irrational manifestations of politics and the pursuit of selfish interests. Thus, political conflict was to be avoided. Planners and policy-makers were to determine the correct courses of action for others to follow and establish rules and procedures that assured compliance with them.

The structural adjustment policies that emerged during the 1980s and early 1990s were based on models that were no less rationalistic. They focused on the functioning and characteristics of national economic systems and used macroeconomic theory and econometric models to prescribe changes in policies. They assumed that structural changes and "getting the prices right" would lead automatically to incentives for private enterprise expansion and widespread economic growth.

Although some officials of international funding organizations would not espouse this explicit interpretation of their views, a highly rationalistic approach was clearly reflected in the methods of planning and management that they have used since the late 1960s. Officials of the World Bank, most bilateral aid institutions, and United Nations agencies have insisted that development projects be identified, prepared, appraised, and selected through comprehensive and systematic analysis. During the 1960s and 1970s they relied heavily on methods and procedures of project management adopted largely from the practices of private corporations engaged in physical construction projects and of government agencies in Western countries concerned with defense systems or space exploration. The methods included cost–benefit analysis, linear programming models, network scheduling, and planning/programming/budgeting systems. Since international lending institutions provided much of the funding for development projects they insisted that their procedures and controls – which they considered to be modern management systems – be adopted as well by bureaucracies in developing countries. Paradoxically, as development strategies changed during the 1970s and 1980s to address more complicated and less controllable problems of human development, procedures for planning and managing projects became even more rigid and routinized (Rondinelli 1976a, 1979a).

The structural adjustment policies of the 1980s and 1990s, while eschewing national comprehensive planning and even the use of projects as lending instruments, were nevertheless based on highly standardized prescriptions and rationalistic models. They called almost universally for fiscal austerity, liberalization of trade, strong monetary controls, rapid privatization of state enterprises, and deregulation of industry, often without an adequate understanding of the social and political characteristics of the countries for which the reforms were prescribed.

Rarely, however, have development policies been carried out in the prescribed ways, and this disparity between theory and reality is at the heart of recurring debates over the effectiveness of conventional methods of development planning and administration (Rondinelli 1987). This book focuses particularly on projects as instruments of policy implementation to examine this disparity. The emphasis is on projects for a number of reasons. During the 1960s and 1970s projects became the primary means through which governments of developing countries translated their plans and policies into programs of action. The rationale was that comprehensive and detailed development plans were of little value unless they could be translated into specific projects that could be designed and implemented efficiently. Thus, projects came to play a central role in the political economy of developing countries. Hirschman (1967: 1) called them "privileged particles of the development process," and Gittinger (1972: 1) saw them at the "cutting edge" of development administration. In theory, projects would promote economic changes by integrating markets, linking productive activities in the public and private sectors, providing the organization and technology for transforming raw materials into economically and socially useful products, and creating physical infrastructure needed to increase exchange and trade (Uphoff and Ilchman 1972). During the 1970s projects were designed to stimulate social change: education, health, family planning and social services projects would help to satisfy basic human needs and provide new skills required in traditional societies to initiate and sustain modernization.

Successful projects would generate new resources for further investment, creating a momentum for widespread and sustained economic growth. In this sense, Hirschman (1967: 1) saw development projects as a special kind of investment: "the term connotes purposefulness, some minimum size, a specific location, the introduction of something qualitatively new, and the expectation that a sequence of further development moves will be set in motion." Because they were considered by their sponsors to be manageable sets of activities, projects became the primary means of translating development policies and aid strategies into programs of action.

A substantial amount of evidence also suggests that translating plans into action was, and continues to be, one of the most difficult tasks facing development administrators. For more than three decades analysts have pointed out that there has been "a scarcity of well prepared projects ready to go and . . . the lack of projects reduces the number of productive investment opportunities" (Waterston 1971: 240). The United Nations Economic Commission for Africa (1969: 137) emphasized that "in many countries essential pieces of development cannot be carried out at the time when they are required, not solely because of financial shortages, but because, the projects themselves have not been technically prepared." Other UN agencies noted (1969: 69) that "there are many cases where the shortage of good projects is even more serious than the shortage of capital or foreign exchange."

Although the World Bank, United Nations Agencies, and many bilateral assistance organizations have given more emphasis to sectoral and program loans during the 1980s and 1990s, many of these large loans and grants have been "projectized" by governments in developing countries, and a large portion of aid is still provided by international agencies through individual projects. In a World Bank report, Cernea (1991: 8) concluded that: "despite the recurrent debates on the merits and disadvantages of projects as instruments of development intervention, no effective alternatives have emerged, and projects are likely to remain a basic means for translating policies into action programs."

For our purposes, the focus on projects affords the best opportunity to illustrate the disjunction between methods of planning and implementation used by governments and international agencies and the nature of development problems. Experience with growth-with-equity and basic human needs strategies during the 1970s showed that they were not amenable to systematic analysis or comprehensive design. The development assistance policies that emerged during the 1970s for alleviating poverty, increasing agricultural productivity, expanding employment opportunities and providing greater access to social services for larger numbers of people made project design more complex and the success of projects more uncertain. As will be seen in Chapter 3, many implementation problems arose from attempts to plan and manage projects aimed at generating social change with techniques and procedures intended for physical facilities and industrial construction projects. Many evaluations indicated, as did an assessment of small-scale agricultural projects in Africa and Latin America, that success requires flexibility in planning and design, opportunity to adjust plans as projects progress, and continuous redesign during implementation. "Few projects can survive a rigid blueprint which fixes at the time of implementation the development approaches, priorities, and mechanisms for achieving success," the evaluators (Morss *et al.* 1975: 329) noted. "Most projects scoring high on success experienced at least one major revision after the project [managers] determined that the original plan was not working." A large degree of flexibility is critical, they argued, "particularly if the technology is uncertain or if the local constraints facing small farmers are not well known."

The argument for greater flexibility and innovation in development administration rests in part on the observation that development policies are complex, uncertain, and require flexible and experimental methods of implementation.

THE TREND TOWARD TECHNOCRACY

The methods of planning, analysis, and management introduced during the 1960s and 1970s – and which still dominate the procedures of most international organizations and governments in developing countries – were not primarily concerned with flexibility, responsiveness, and learning. They were more concerned with efficiency and control. Rationalistic approaches were introduced not only because they were compatible with economic growth strategies that were prevalent during the 1950s and 1960s, but also because they were perceived to be effective methods of reducing uncertainty and increasing efficiency. Administrators embraced control-oriented planning and management techniques that were either ineffective or inherently incapable of reducing uncertainty at the same time that development policies were becoming more complex.

The adoption of rationalistic approaches to planning and management, as Benveniste (1972: 27) points out, was also quite consistent with the "conventional notion that planners and government need strong power to plan; that planning has to be 'imperative' in nature; and that short of centralization and strong executive control to impose the plan, it will fail." It was reinforced by the growing realization that despite the preparation of national development plans governments in developing countries controlled few of the major factors affecting economic and social change. In many cases rationalistic approaches to planning and management were related "to the inability of organizations or entire governments to function in an environment that has become too uncertain" (Benveniste 1972: 24). Moreover, the trend toward quantification and the introduction of systems analysis wrapped national planners and foreign experts in the mantle of authoritativeness. Their tools became their power. "The expert's dependency on measurement is very real. Measurements and quantitative analysis are the bases of the knowledge which differentiates them and, therefore, a basis of their social power," Benveniste (p. 57) noted. "They cannot spend too much time talking in vague ways. Sooner or later they need to concentrate on the issues on which they can exercise their skills."

In addition, the adoption of rationalistic and quantitative methods of planning suited the needs of politicians in many developing countries, for such methods offered a variety of benefits that were not directly related to their ostensible function of selecting economically feasible and optimal courses of action (Benveniste 1972, Davis 1974, Rondinelli 1975). Rationalistic analysis was used to legitimize political decisions that had already been made or to justify not taking any action at all to resolve pressing political issues. It was used to justify raising the priority of some political options that were not popular and removing others from the policy agenda. It was a means of making fundamentally political decisions seem objective and technical and of altering the time horizon for decision-making. Often the mere attempt to do such analysis was a symbol of rationality and objectivity. Central planning and control was a means of constraining the process of political interaction in decision-making and enhancing the influence of the bureaucracy (Rondinelli 1987). Finally, the procedures often served as "window-dressing" to meet the requirements of international agencies and national finance ministries for systematic analysis and rational planning.

CHANGING PERCEPTIONS OF OPTIMUM DEVELOPMENT STRATEGIES

The directions of international development policy shifted drastically in the early 1970s from a concern with promoting rapid growth in gross national product through capital-intensive industrialization, export production and construction of large-scale physical infrastructure to ways of stimulating internal demand, expanding economic participation, developing human resources and reducing disparities in income and wealth. The new policies were as concerned with the composition and distribution of the benefits of development as with the rate of economic growth.

Changes in development policies during the 1970s resulted from a number of converging forces. In part, they evolved from the realization that macroeconomic development strategies after the Second World War had not been effective; the trickle-down and spread effects promised by conventional economic theory had not materialized in much of the developing world and could no longer be used as the basis for planning. Nearly half of the 187 nations surveyed by the World Bank in 1974 had a per capita GNP below $500 a year, and only 19 reported levels above $5,000. The majority of countries with a per capita GNP below $300 during the 1960s also had a relatively low annual growth rate. About 30 desperately poor countries with intransigent social and economic problems, and with a substantial portion of the world's population, had been bypassed entirely by economic progress. Moreover, the distribution of income and wealth, even within growing economies, remained highly inequitable. In the mid-1970s the World Bank estimated that nearly 85 percent, or more than 750 million, of the people in developing nations were living in relative poverty, earning less than $75 a year; two-fifths subsisted in absolute poverty on annual incomes of less than $50 (World Bank 1975).

New strategies also emerged from internal social and political pressures, from widespread recognition of the social injustices perpetuated by dual economies, and from the realization by some national leaders that political stability and national unity could not be achieved or maintained with only narrow support from a small, wealthy, urban elite. Moreover, the seeming success of experiments with more equitable growth policies in some communist and socialist societies created pressures on governments with mixed economies to reduce the growing gaps in income and wealth.

International evaluation commissions frequently criticized development policies and aid strategies as inadequate or inappropriate. The Pearson Commission (1969) noted that growth in the economies of many poor nations continued to be outpaced by population increases, thereby "nullifying much of the development effort." The successful implementation of some development policies simply produced new and more difficult problems. As massive educational development programs began to reduce illiteracy and ignorance, frustrations were created for millions of young graduates who could not find jobs in stagnant economies. As crop yields and food supplies increased with the success of agricultural innovations, governments in many developing countries had neither the resources nor the technical capabilities to cope with the resulting problems of production, distribution, and marketing. Even modest economic progress generated new expectations that governments could not fulfill. "In supplying such facilities as schools, hospitals and urban housing," the Pearson Commission (1969: 12–13) reported, "even a holding operation to keep standards from slipping is often difficult."

When it became clear in the late 1960s that capital-intensive industrialization policies had not been effective in promoting growth and alleviating poverty in most developing countries, new strategies and goals appeared in national development plans. India's plan for the 1970s admitted explicitly that the benefits of aggregate economic growth had not filtered down to the vast majority of the people and that without more widespread participation in economic activities growth could not be accelerated or sustained. "Growthmanship which results in undivided attention to the maximization of GNP can be dangerous," Indian analysts noted. "Elimination of abject poverty will not be attained as a corollary to a certain acceleration of the growth of the economy alone" (Das 1973: 430). Policy pronouncements in other countries came to similar conclusions. In its 1974–8 national plan the Philippine government proclaimed "that no longer is maximum economic growth at the singular apex of goals; equally desirable are maximum employment, promotion of social development and more equitable distribution of income and wealth." Thailand's plans for the 1970s sought to balance national development more equally among its geographical regions and restructure the economy to reduce income inequalities, develop human resources, and expand rural employment. Development plans in many African and Middle Eastern countries began to stress the importance of employment generation and wider distribution of income (Nigam 1975).

Bilateral and international assistance organizations also modified their aid strategies in the early 1970s. In 1973, for instance, the United States Congress issued a new mandate to the US Agency for International Development to give the highest priority to activities in developing nations that "directly improve the lives of the poorest of their people and their capacity to participate in the development of their countries." In the Foreign Assistance Act of 1973, Congress declared that the conditions under which American foreign aid had been provided in the past had changed, and that policy should also be changed to reflect the "new realities."

Although American aid had successfully stimulated economic growth and industrial output in some countries, the House Committee on Foreign Affairs lamented that the gains "have not been adequately or equitably distributed to the poor majority in those countries," and that massive social and economic problems prevented the large majority of people in other nations from breaking out of the "vicious circle of poverty which plagues most developing countries" (*US Code Cong. and Admin. News* 1973: 2811). The Act asserted that henceforth American aid would depend less on large-scale capital transfers for physical infrastructure and industrial expansion, and more on transferring technical expertise, modest financial assistance and agricultural and industrial goods to meet "critical development problems." Aid would be used to promote changes that affect the lives of the majority of people in developing countries: in food production, rural development, nutritional standards, population planning, health and educational services, public administration, and human resource development.

The World Bank was also reordering its priorities and procedures. In his address to the Board of Governors in 1973, World Bank President Robert S. McNamara noted that the benefits of international aid had not filtered down to the majority of the poor, and insisted that the Bank increase investments in those sectors and areas of the world where they would have the greatest impact on raising the productivity and income of "nearly 800 million individuals – 40 percent out of a total of 2 billion – [who] survive on income estimated at 30 cents a day and in conditions of malnutrition, illiteracy and squalor." He challenged the Bank to design "new style" projects for achieving growth-with-equity, and called for rapid increases in the amount of loans for agricultural and social projects made to the Bank's poorest members. He placed a new emphasis on loans for multi-purpose, integrated, low-cost, replicable projects designed to benefit the poorest groups by increasing the productive capacity of small-scale farmers and rural industries. McNamara (1973: 10) called for a comprehensive program for alleviating poverty in rural areas. Such a program would require, among other things, more rapid land and tenancy reform, better access for the poor to credit, expansion of potable water supplies, extension of technical assistance to subsistence farmers, greater access by the poor to public services, "and most critical of all, new forms of rural institutions and organizations that will give as much attention to promoting the inherent potential and productivity of the poor as is generally given to protecting the power of the privileged."

Similarly, UN specialized agencies, such as the International Labor Organization (ILO), began calling for new employment-generating strategies that would help the world's poor to satisfy their "basic human needs." In its Declaration of Principles, the World Employment Conference organized by ILO in 1976 argued that "past development strategies in most developing countries have not led to the eradication of poverty and unemployment," and that major shifts were needed in international development strategies to ensure full employment and adequate incomes for the majority of the poor in the developing world as quickly as possible. Development plans and assistance policies, it asserted, should focus on providing for the minimum private consumption and community service needs of the poor and on promoting full employment as a way of generating the income required by the poor to satisfy those needs (ILO 1976: 5–6).

Previous assistance policies were attacked not only for their ineffectiveness in promoting economic growth but also because they reinforced those arrangements in the international economic system that worked to the disadvantage of poor countries. Development assistance was seen by many Third World political leaders as an extension of colonialism and imperialism, designed to attain the economic and political objectives of rich countries rather than to alleviate poverty or promote growth in developing nations.

In 1980, the Brandt Commission warned of dire consequences for the future of mankind unless a new international economic order could be forged to reduce the growing gap in income and wealth between rich and poor societies (Independent Commission on International Development 1980). But drastic changes in the international economy at the beginning of the 1980s illustrated more forcefully the degree of uncertainty and complexity that was inherent in the development process. International economic recession raised serious doubts about the ability of international organizations and governments in developing countries to promote or sustain high rates of economic growth and created formidable financial obstacles to reducing poverty. Deep recession and slow growth in western industrial economies had world-wide effects – raising interest rates on investment capital, reducing demand for the exports of developing countries, and increasing the costs of their imports. The rapidly rising price of petroleum – an 80 percent increase between 1978 and 1980 alone – had devastating effects on the economies of oil-importing countries, which included most of the world's poorest nations. Debt service rose to unprecedented levels in many developing countries, placing strains on already meager development budgets and threatening the survival of many of the projects and programs started in the late 1970s. International financial assistance to developing nations decreased sharply at a time when the need for additional resources was becoming more urgent (World Bank 1981).

World Bank officials predicted that even under optimistic international economic conditions the income gap between the richest and poorest countries would continue to grow, and that if pessimistic projections prevailed the number of individuals living in absolute poverty would continue to rise. Again, trends in international assistance policy and development strategy changed. The World Bank placed stronger emphasis on reforming the economic structures of developing countries. The concern with alleviating poverty and raising the incomes and living conditions of the poorest groups began to wane. Many of the human resource development programs that were found to be essential to sustained economic growth during the 1970s were labelled welfare policies by conservative economists who gained prominence in international assistance organizations during the early 1980s. Governments in developing countries had to make rapid, and sometimes traumatic, economic reforms to maintain their access to international sources of capital.

The reasons for this change in the focus of development assistance policy are not difficult to discern. The rate of economic growth in developing countries dropped from an average of 5.4 percent a year during 1973–80 to 3.9 percent during 1980–7. Average annual growth in per capita GNP dropped from about 3.9 percent in the late 1960s and early 1970s to about 1.7 percent during the 1980s.

Developing countries" indebtedness increased from $49 billion to almost $775 billion between 1970 and 1986 and the cost of debt service increased from 10 to 20 percent of the value of exports (World Bank 1988). The long economic recession of the late 1970s and early 1980s, accompanied by high rates of inflation and large budget deficits, weakened the economies of many developing nations.

In the hope of reversing these trends, the International Monetary Fund (IMF) and the World Bank turned to two types of policy reforms: economic stabilization and structural adjustment. *Economic stabilization policies* – usually requiring economic austerity measures – aimed at controlling inflation, the loss of foreign exchange reserves, capital flight and public sector deficits. *Structural adjustment policies* sought to remove obstacles to long-term economic growth by promoting economic liberalization, eliminating excessive taxes and subsidies, controlling prices and interest rates, reducing high tariffs and import restrictions, and modifying or eliminating distortions in incentives for private sector investment. The World Bank's structural adjustment loans also sought to reduce government employment and public expenditures, privatize public services, and reduce government interference in market activities (Lal 1987, Khan 1987).

Several international organizations, including USAID and the World Bank, prescribed public choice theory as a means of assessing the feasibility of privatizing the provision of public services. Public choice theory – the application of economic principles to assessing the demand for and supply of publicly provided goods and services – relied heavily on analyzing the aggregation of individual preferences through political coalitions and market mechanisms, and promoted the provision of services through private institutions.

CHANGING POLICIES AND CONVENTIONAL PERSPECTIVES ON PLANNING AND IMPLEMENTATION

Although these changes in development policies and aid strategies were strongly influenced by rapidly changing international economic conditions and evolving theories of economic development, they failed to reflect lessons about the nature of development problems and never seriously re-examined their assumptions about how policy planning and implementation should be done.

The methods of planning and administration that governments in developing countries and international assistance organizations employed have nearly always been inappropriate to the nature of development problems. They more often obstructed or constrained rather than encouraged experimentation and social learning. Leaders of developing countries and officials of international assistance organizations have more readily acknowledged the complexity and uncertainty of development problems in recent years, and modified their policies and programs as they learned from experience. But their analytical and administrative procedures have yet to come to grips with the experimental nature of development policies and with the uncertainties of the development process.

In a critique of the Brandt Commission report issued in 1980, former World Bank official P.D. Henderson (1980: 16) highlighted the discrepancies between prescription and reality that plagued nearly all policy statements on international development problems. He pointed out that the Brandt report "embodies a high degree of confidence that the future either is or can be predictable and controllable". He noted that "too little account is taken of the pervasive and irreducible uncertainty which surrounds the future, nor would one gather from the Report that much of the uncertainty that could in principle be reduced arises from the behavior of governments."

Throughout the 1970s and 1980s, international assistance organizations continued to assume that their policies, programs and projects could be rationally conceived and comprehensively planned solutions to development problems. Henderson's reaction to the Brandt Commission report was one of the few to challenge its assumptions about the certainty of development problems and to question whether:

> the issues of social and economic life are such that it makes sense to think in terms of "solutions" to them – as though they were like the entries in a crossword puzzle, for which there can be found a recognized, uniquely correct, and permanently valid set of responses.
>
> (Henderson 1980: 16)

In reality, development planners and officials of international lending institutions have attempted to promote social and economic reforms in developing societies with relatively little knowledge of the conditions they were seeking to transform and with little certainty that their theories, policies and projects would produce the desired effects. Indeed, neither the array of factors causing poverty and underdevelopment nor the dynamics of growth in the international economy have ever been fully understood.

The only certainty to emerge from past experience is that development problems are extremely complex, differing drastically among societies and over time. Policies pursued successfully in one country did not necessarily work in others, and conditions that promoted or obstructed change in some societies did not prevail in all. The export industrialization policies that successfully promoted economic growth in Taiwan and South Korea during the 1960s and early 1970s, for instance, could not be replicated later in Latin American or African countries. China's programs for providing basic human needs and promoting a more equitable distribution of income during the 1950s and 1960s could not launch the country into a more dynamic process of economic and political development during the 1970s and 1980s, and did not even achieve similar results when they were tried in other socialist countries. Political and administrative obstacles undermined the implementation of structural adjustment policies in many developing countries during the 1980s and 1990s, and some governments failed to carry out economic stabilization programs (Nelson 1984). The diverse values, ideologies, attitudes and behavior of people and governments in developing nations have made it impossible to define development objectives universally. And although some economic and social problems were found to be common to many poor countries they also differed in significant ways, making general solutions inappropriate to any particular nation. Even global influences such as inflation, rising prices for factors of production, and changing patterns of international trade affected different countries in different ways.

The success of economic development policies in the future is likely to depend on the resolution of two basic issues. One is how developing nations can mobilize and use their own resources more effectively to increase the productivity and income of millions of people living in poverty. The other is how international assistance organizations, governments and the private sector can organize their planning and administrative procedures to cope more effectively with the growing complexity and uncertainty of development problems. The two issues are related and both are examined in this book.

The focus of the book, however, is less on the substance of policies and more on the processes and methods by which development programs and projects are formulated and implemented. The disjunction between the nature of the development process and the methods of project planning and management used by governments and international agencies has been an important factor accounting for the disappointing results of development assistance in the past, and its impact will be even more serious in the future.

PUBLIC POLICY MAKING AS SOCIAL EXPERIMENTATION

The experimental nature of development policies becomes more evident when the history of development theories and aid policies is viewed, as it is in Chapter 2, from a perspective different from that underlying conventional development theory and when public policy-making is recognized as an incremental process of trial and error through political interaction and successive approximation (Lindblom 1965, Wildavsky 1979, Rondinelli 1987). As Johnston and Clark point out, in their review of rural development programs in developing countries, the commonly held perception of development problems differ substantially from reality:

> To speak of the "development problem" is to imply a well structured world of unambiguous objectives, mutually exclusive choices, authoritative decision-makers and willing decision-endurers. In contrast to this idyllic vision, development actually involves a staggering variety of people and organizations all pulling and pushing and otherwise interacting with each other in pursuit of their various interests. Since each actor in this process has a more or less unique perspective on "the development problem," the policy process invariably deals with numerous overlapping problems, or in Ackoff's phrase, a "mess." Turning messes into problems about which something constructive can be done is one way of viewing the central task of policy analysis.
>
> (Johnston and Clark 1982: 11)

Such perceptions of the dynamics of development suggest a far different approach to planning and management than is usually prescribed by those who hold a rationalistic or technocratic view of problem solving. "The planning model recognizes that implementation may fail because the original plan was infeasible. But it does not recognize the important point that many – perhaps most – constraints remain hidden in the planning stage, and are only discovered in the implementation process," Majone and Wildavsky (1978: 106) point out. This view implies that few programs and projects can be designed in great detail and carried out in highly coordinated and controlled fashion in accordance with comprehensive plans. "When problems are puzzles for which unique solutions exist, technicians can take over," Majone and Wildavsky contend (p. 113). "But when problems are defined through the process of attempting to draft acceptable solutions, then analysts become creators as well as implementors of policy." Because the ability of development planners to predict and control the outcomes of their programs and projects under conditions of uncertainty is quite limited, their methods must be better suited to recognizing and dealing with uncertainty, detecting and correcting errors, generating and using knowledge as experiments progress, and modifying actions as opportunities and constraints appear during implementation. It will be seen later that few of the planning and management techniques now used by governments and international assistance agencies meet these criteria. Instead, many suppress social learning and constrain experimentation.

DESIGNING AND ORGANIZING PROJECTS AS POLICY EXPERIMENTS

It is the uncertainties and complexities of development policies and responsiveness to the needs and capabilities of intended beneficiaries that must become the major concern of those providing financial and technical assistance. Although little can – or should – be done to try to uncover and control them entirely at the outset of a project, the uncertainty and ignorance involved in pursuing policies should be realistically assessed. Attempts to plan in more detailed and precise fashion should proceed incrementally, as uncertainties or unknowns are reduced or clarified during implementation. Planning must be viewed as an incremental process that tests propositions about the most effective means of coping with social problems, reassessing and redefining both the problems and the components of development projects as more is learned about their complexities and about the economic, social, and political factors affecting the outcome of proposed courses of action. Complex social experiments can be partially guided but never fully controlled; thus, analysis and management procedures must be flexible and incremental, facilitating social interaction so that those groups most directly affected by a problem can search for and pursue mutually acceptable objectives. Rather than providing a blueprint for action, planning should facilitate continuous learning and interaction, allowing policy-makers and managers to readjust and modify programs and projects as they learn more about the conditions with which they are trying to cope (Korten 1980, Korten and Alfonso 1981). Planning and implementation must be regarded as mutually dependent activities that refine and improve each other over time, rather than as separate functions.

If, in fact, all development activities are essentially experimental, fundamental changes are needed in the way governments in developing countries and international assistance organizations formulate development policies and implement projects.

First, it is highly unlikely that uncertain, experimental activities can be planned comprehensively and designed in great detail at the outset. As will be seen in Chapters 2 and 3, a variety of economic, social and political forces operate in poor countries to make detailed planning difficult or impossible. The lack of knowledge about the groups and societies for which assistance projects are proposed makes detailed design ineffective or perverse. The inability to control processes of interaction that affect the success of projects makes attempts at comprehensive planning costly and inefficient, and the pressures to make administrators and managers conform to preconceived designs discourage them from taking appropriate action to deal with unanticipated difficulties as they arise.

The experience in developing countries described in Chapter 4 indicates that planners and administrators have had as much difficulty applying quantitative, systematic, and comprehensive techniques of analysis, planning, and management as those in western industrial countries. The growing criticisms of cost–benefit analysis, planning/programming/budgeting systems, critical path methods of scheduling, quantitative model building, and management systems, when they are applied to social or political problems in western countries, are based not only on their inappropriateness but also on their ineffectiveness. The dangers of transferring them are exacerbated by the reluctance of experts to inform planners and administrators in poor countries of their limitations. A long list of criticisms of quantitative analysis and systems management has been compiled by social scientists in western industrial countries (Wildavsky 1969, Schick 1973, Sapolski 1972, Hoos 1972). Among the most important criticisms are:

1 Systems analyses require a concise definition of goals and objectives, but the goals of social programs and policies often cannot be clearly defined because they are really expressions of

social values on which various groups usually disagree and which cannot be stated in purely economic terms;

2 There are severe problems in quantifying and measuring the results or effects of public programs because there is often no physical product associated with them;
3 Identifying and categorizing inputs, outputs, costs and benefits is a subjective rather than an objective process on which analysts and interest groups often disagree;
4 Government agencies at all levels, even in rich countries, lack the administrative capacity and analytical talent to do these types of analyses effectively;
5 Obtaining adequate and reliable data on which to base systems analysis and synoptic planning is difficult at best and often impossible, especially within bureaucracies in which administrators view such analyses as a threat to their political influence;
6 Systems analyses often ignore or discount the complex processes of social interaction through which decisions are really made, processes that are based on decision-making through bargaining, exchange and compromise;
7 Political leaders often either do not understand or are unwilling to accept the results of sophisticated quantitative analyses and therefore tend to ignore them in arriving at decisions;
8 Because the programs or policy options subjected to systems analysis usually differ in their objectives, characteristics, purposes, and content, it is difficult or impossible to make relevant comparisons and tradeoffs among them or to deal with the "crossover" problems of treating as coherent programs similar activities that are the responsibility of different agencies and organizations;
9 These methods are often ineffective because systems analysis and comprehensive planning are episodic and time consuming, whereas political decision-making is continuous or cyclical and requires immediate and continuing supplies of information; and
10 Systems analyses usually address questions of how to maximize the utilities of individual decision-makers whereas policymaking is primarily concerned with questions of how to distribute public resources to groups with different goals and purposes in the most politically effective way.

These criticisms of rationalistic methods, and especially of cost–benefit analysis, in rich countries thus arose not only from the great difficulty in operationalizing them when they were applied to development problems, but also because they were incompatible with the way political decisions were made. Moreover, they discouraged analysts from understanding the complexity and uncertainty of the problems with which they were dealing. Meltsner (1976: 32) noted in his study of policy analysts in the United States that they were ineffective precisely because they tried to clarify and make technical those issues that were inherently complex and political. "What the technician shuns, the politician embraces," he pointed out. Benveniste argued that the requirements of systems analysis are often irreconcilable with political requirements:

> Goal specification requires a high level of political consensus. When social consensus exists goal specificity does not create political costs. When many divergent views exist, however, the possibility of establishing well-defined goals that satisfy everyone becomes much more difficult. Even the process of spelling out goals may result in considerable conflict as each contending faction struggles to place its own preferences high on the list of objectives. Vague and ill-defined goals are an equivalent to having secret goals. As long as goals are secret, it is possible for competing groups to pursue their own ends without necessarily encroaching on each other.
>
> (Benveniste 1972: 70)

Many of the rationalistic approaches to planning and analysis prescribed for developing countries during the late 1980s and early 1990s were also the subject of intense debate in the countries where they originated. Critics argue that the neoclassical economic analyses that were used to formulate structural adjustment policies and the application of public choice theories both made assumptions that were not always realistic in developing countries. Public choice theories were developed in the United States where it was assumed that a high degree of physical, economic and social mobility allowed people to move to communities where desired "market baskets" of goods and services are provided; that income levels were sufficient to allow people to purchase the market basket of goods and services they desired; and that information about price and quality of services was available and widely shared among potential consumers. Most of these conditions were not characteristic of poor countries, a major limitation of relying on public choice theory to analyze the feasibility of privatizing public services.

Many developing countries had only weak markets with imperfections that undermined or weakened the very conditions assumed by public choice theorists.

Institutional economists were critical of the public choice theorists" assumption that market transactions were the only effective means of providing services and of neoclassical economists" tendency to ignore institutional factors in public policy-making (Langois 1986; Nelson and Winter 1982). Public choice theories were rarely concerned with improving the capacity of government agencies to provide more efficiently collective goods that could not be provided through market mechanisms, a primary concern of governments in many developing countries where the private market was still weak or embryonic. Public choice theories were based on the assumption that people acted rationally, always pursued their own economic self-interests, and would make optimal economic choices if left unfettered by government regulations and constraints.

To the extent that structural adjustment policies were formulated from macroeconomic models, they suffered from many of the limitations of that form of analysis. Kamarck (1983) has pointed out that economic models that attempt to cover a whole economy inevitably suffer from "political-economic" errors: that is, the failure to recognize that many economic issues are also political issues and that decisions about them are made by criteria that cannot easily be captured by mathematical equations. The more comprehensive the economic model becomes, the greater the margins of error in the data and in the knowledge and sharpness of the concepts that are embodied in the parameters. The model-builders must often make highly questionable assumptions about behavior, which is usually considered to be invariant, and to ignore or discount many exogenous variables. They usually fail to take into consideration how cultural, social, and political differences affect human behavior. More important, they often have to assume that future changes will be extensions of past trends and thus they cannot easily predict major disjunctions. "It would indeed be surprising, therefore, if such macroeconomic models did succeed in correctly predicting major changes in the economy that were not obvious to anyone with informed common sense," Kamarck (1983: 71) concluded. "And, in fact, the record of performance substantiates this skepticism."

Other critics have questioned whether in fact any of the rationalistic approaches have led to a more reflective and systematic process of decision-making. Peter Self (1975: 91–2) questioned both the logic and the assumptions of what he called the myth of numerology, or the belief that "rationality equals counting and measuring." He questioned whether quantification necessarily causes policy-makers to think more carefully or broadly about the factors influencing the outcome of a decision, or if it is even a necessary condition for deliberation and reflection. "The critical question here is how carefully and widely an individual is prepared to think about his decision premises," Self contended. "If he thinks in a narrow way numerology will not make him reflective, and if he is reflective there is no need for numerology." The greater danger is that rationalistic approaches encourage analysts to think more narrowly, to consider only what is quantifiable and to reduce uncertain and complex conditions to abstract or unrealistic assumptions merely to fit a deductive model. Moreover, Self (p. 93) argued that in cost–benefit analyses the "fixing of eyes upon the techniques of final evaluation shifts attention from the intervening process and the reasoning behind them. The conclusions become numerate but shallow."

The point was underlined by former US Secretary of Defense James Schlesinger when he noted that the availability of analytical tools "may obscure the unpleasant reality that many public policy problems are highly intractable." The real problem, he argued:

> is not method but understanding; the linking of costs and benefits is only as good as man's knowledge. Yet there exists an image of systems analysis in which expert practitioners apply their methods and grind out solutions to policy problems – in the absence of a deep knowledge of the relevant social institutions or real world mechanics.
>
> (Schlesinger 1968: 282)

In their review of experience with rural development in Third World countries, Johnston and Clark (1982: 231) argued that cost–benefit analysis and quantitative models have been applicable to a very small number of routine and tactical problems and that they have proven to be "of little value in guiding the strategic debate over, for example, the priority that should be given to nutrition, health and family planning programs or the choice between unimodal and bimodal agricultural development." Bromley and Bustelo (1982) emphasized the dangers of transferring such techniques to developing countries in Latin America without providing planners and administrators with adequate information about their limitations and inappropriate applications. They complained that much of the technocratic methodology was introduced in Latin America simply to satisfy the requirements of international aid agencies and not from local perceptions of its usefulness. Prescriptions for the adoption of rationalistic analysis and management systems continued because international development banks and powerful groups of national consultants supported and required them, "enabling technocratic firms to make substantial profits by selling their sophisticated expertise and their knowledge of the latest techniques."

But even if such techniques could be made operational, the complex and uncertain nature of development makes it impossible to anticipate, analyze, and control all of the variables that affect development projects. Hirschman was correct in his conclusion that if planners of most of the World Bank projects that he observed had been able to anticipate all of the problems that later befell them, few of the projects would have been judged to be feasible and actually undertaken. Yet, creative managers were often able to cope with unanticipated obstacles by redesigning projects as they progressed, and to accommodate to constraints and opportunities as they arose. "The term 'implementation' understates the complexity of the task of carrying out projects that are affected by a high degree of initial ignorance and uncertainty," Hirschman (1967: 35) contends. "Here project implementation may often mean in fact a long voyage of discovery in the most varied domains, from technology to politics."

If that voyage is to be successful, the methods used to formulate and carry out development policies must encourage the process of discovery and use information and experience to chart a course of action along the way. This implies that planning and implementation must be more closely integrated and that project planning must proceed through a series of stages in order to reduce uncertainties and unknowns: from highly experimental activities that probe possible courses of action to pilot and demonstration projects that test alternatives and identify conditions under which interventions are more or less effective. Some experimental and pilot projects may lead to full-scale replication. Others may be useful only under particular conditions and not replicated, or found to be inappropriate or ineffective, and therefore abandoned.

As will be seen in Chapter 5, *experimental projects* are needed when little is known about problems or the most effective means of setting objectives. Unknowns and uncertainties affect nearly all aspects of development projects, from the definition of problems and the feasibility of alternative interventions to the choice of the most appropriate technologies and organizational arrangements. Although experimentation is quite common in agricultural and health projects, much less attention has been given to experiments in other economic and social projects. In most projects that are sponsored by governments and international agencies little is initially known about the conditions or needs of the intended beneficiaries, their attitudes and behavior or the social and political environment in which projects must be implemented. Nor is much known about appropriate types, magnitudes and combinations of resources needed to deal with development problems in many countries, the most effective timing and sequencing of activities, or the best production and delivery systems for intended beneficiaries. International agencies and governments of developing countries have relatively little experience with designing and carrying out development projects as experimental activities, but the little they do have suggests that such projects should be of moderate scale. They often can be carried out by beneficiaries, but they frequently require technically trained staff. In any case, they must elicit the participation of potential beneficiaries in the design and organizational phases, their managers must have easy access to specialized inputs and resources, and methods of implementation must be flexible (Hapgood 1965, Lele 1975). Experimental projects must often be isolated or sheltered from routine administrative responsibilities and from political pressures to replicate them too soon or to abandon them too early. In Chapter 5 it will be argued that conventional forms of scientific experimentation are not appropriate for most development projects and that other, less formal, processes of social experimentation must be tried.

Pilot projects can be used to test the results of experiments under a greater variety of conditions and to adapt and modify to local conditions and needs the methods, technologies, or procedures that have proven to be effective in other countries. It will be seen in Chapter 5 that pilot projects are most appropriate when the problem or objective of a policy is well defined or when much is already known about the effects of small-scale experiments. Pilot projects are usually intended to test new methods and technology, to determine their relevance, transferability and acceptability, and to explore alternative ways of disseminating results or of delivering goods and services. The relatively few studies that have been done of pilot projects suggest that careful attention must be given to choosing appropriate locations, structuring activities to fit local needs and conditions, collecting baseline data, and, especially, monitoring and evaluating the project to determine conditions that influence success or failure. The projects must be evaluated in such a way that preconditions for replication and successful demonstration can also be determined. Pilot projects must also be organized to shelter their managers from undue political pressures to show quick results, since implementation is still primarily by trial and error, requiring creative and flexible administration.

When pilot phases have been completed, *demonstration projects* can be used to exhibit the effectiveness and to increase the acceptability of new methods, techniques or forms of social interaction on a broader scale.

Although demonstration projects may be less risky and uncertain than experimental and pilot projects, innovative and creative management is still required to gain acceptance of new ways of doing things. Experience with agricultural, family planning, rural development, and small-scale industrial projects in Africa and Latin America clearly indicates that successful demonstrations must be profitable to participants as well as novel and innovative, yet related to their experience. Demonstration projects should include all components required to support successful adaptations, be relatively simple to understand, and make use of accessible materials and tools. Moreover, they must be highly visible, produce results that are easy to explain, and be replicable with a minimum of supervision and training (Hapgood 1965, Shaefer-Kehnert 1977, Rondinelli 1979b).

Replication, diffusion or production projects should evolve from experimental, pilot, and demonstration phases. Widespread replication can be undertaken when some of the uncertainties and unknowns have been dealt with, or when a great deal is already known about the elements or potential impacts of a project that would ordinarily be tested in experimental, pilot, and demonstration activities. A primary concern in planning and implementing these projects is to test full-scale production technology and to organize an effective delivery system for disseminating results or distributing outputs.

It must be kept in mind, however, that all development projects are somewhat experimental and that even seemingly routine replications often meet unanticipated difficulties when projects are transferred from one culture to another.

The perception of development policies as experimental activities implies that new forms of analysis, planning, and administration must be devised that are better suited to learning and to uncovering and coping with the uncertainties and risks in policy implementation. An adaptive approach to development administration is outlined in Chapter 6. The greater the degree of uncertainty and the larger the number of unknowns involved in a proposed course of action, the more flexible and probing the methods of analysis – and the more adjustable the procedures of imple- mentation – must be. In Chapter 6 it is argued that the objectives of planning should not be to control in fine detail the activities that will be pursued during implementation, but to increase the opportunities for those managing a project to take appropriate action as they learn more about the conditions affecting their activities.

Indeed, the perception of development assistance offered in the following chapters suggests that a primary purpose of projects should be to build up gradually the planning and administrative capabilities of people and organizations in poor countries rather than simply spending larger amounts of money to build yet another highway or factory for them. By designing and organizing projects to reduce uncertainties and unknowns incrementally, integrate planning and implementation, and use the acquired knowledge to alter and modify courses of action during project implementation, they will become more effective instruments of learning and make greater contributions to development in the future.

2

DEVELOPMENT POLICIES AS SOCIAL EXPERIMENTS

From macroeconomic growth to sectoral development

A review of the evolution of economic development theories and aid strategies since the 1940s yields three propositions that support the arguments set out in Chapter 1: first, development theories have evolved from a complex process of social experimentation and political interaction; second, the methods of planning and implementation used by governments and international assistance organizations have rarely been appropriate for dealing with the complexities and uncertainties inherent in development problems; and third, the development strategies used by governments and international organizations rarely succeeded in strengthening the institutional capabilities of developing countries to sustain the benefits of external assistance.

The complex and uncertain changes in development strategies can be seen in three major periods in the recent history of development assistance. First, during the 1950s, international assistance organizations prescribed industrial development policies that would stimulate maximum growth in the economies of developing nations and assumed that trickle-down and spread effects from increasing national production would automatically alleviate poverty. These policies sought rapid growth in national output with little concern for distributive effects, and thus relied on largely untargeted aid programs.

During the 1960s some attempts were made to compensate for the lack of spread effects. Development assistance strategies were designed to overcome obstacles and eliminate bottlenecks to economic growth by redistributing productive assets, developing human resources, controlling population growth, and increasing productive capacity in lagging sectors of developing economies. Sectoral development loans sought to change those social and economic conditions that were considered to be obstacles to development. These sectoral development strategies used semi-targeted aid: international organizations focused their technical and financial assistance on sectoral development and on changing the behavior of people in developing countries with characteristics thought to be adverse to economic growth.

During the 1970s, changes in the policies of international organizations seeking economic growth with greater social equity marked a second identifiable period in the evolution of development theory. Governments in developing countries and international organizations expressed as much concern with the distribution of benefits as with the rate and pace of economic output. International organizations sought to channel aid to the poor majority (primarily to subsistence populations in rural areas), provide for basic human needs in the poorest countries, and improve the living standards of targeted groups of the poor.

A third era in development assistance strategy evolved from the turbulent changes in the world economy and in the economic, social, and political conditions within developing nations during the late 1970s and early 1980s. In an environment of greater economic uncertainty, the objectives and approaches to foreign assistance changed quickly. During the 1980s and early 1990s, international development organizations, led by the World Bank, shifted their assistance strategies to restructuring the economies of developing countries by providing adjustment loans aimed at increasing export production, liberalizing trade, decentralizing government, and privatizing state-owned enterprises (Rondinelli and Montgomery 1990). They gave higher priority to increasing private sector productivity and less to meeting the basic needs of the poor.

This chapter reviews the development policies and methods of development planning and administration of the 1950s and 1960s; and the following chapter examines changes in development strategies from the 1970s to the early 1990s.

GROWTH MAXIMIZATION AND TRICKLE-DOWN STRATEGIES

When the Point Four Program, or Marshall Plan, was proposed in the 1940s, the intention of the United States and its allies was to rebuild the physical and industrial structure of European countries that had attained relatively high levels of productive capacity prior to the Second World War. International aid programs sought to rehabilitate physical infrastructure and industrial plant, temporarily feed large numbers of people displaced from their jobs by the war, and re-establish market systems in Europe (Hogan 1987). Other international aid programs established in the wake of the Marshall Plan had similar objectives. The World Bank's mission, for instance, was clearly reflected in the organization's formal title – the International Bank for Reconstruction and Development. Concern for promoting development in poor countries was subordinate to reconstructing productive capacity in more economically advanced nations that had been devastated in a long and intense global conflict. In the late 1940s and early 1950s the emphasis of aid-giving organizations was on macroeconomic development, national planning, construction of capital-intensive industries, highways and power-generating systems, and on rebuilding the financial capacity of European countries to invest in their own reconstruction. The plans called for large-scale and expensive projects requiring sophisticated engineering skills and high technology equipment. The governments receiving aid were generally experienced in industrial development and had well-trained professionals and skilled workers, high levels of planning and managerial capability, and a strong motivation to recover as quickly as possible (Mikesell 1968, Mason and Asher 1973).

As they achieved success in rehabilitating the economies of Europe, bilateral and international aid organizations increasingly turned their attention to poorer nations of the world that had never attained high levels of industrial production. The economic, social, and political characteristics of these poor countries were quite different from those of European nations, as were the motivations of their leaders and the extent of poverty within their populations. Preconditions for economic growth that were taken for granted in European nations did not exist in most countries of the Third World.

But international aid agencies pursued much the same strategies in poor countries as they had used successfully in reconstructing the economies of Europe. Little attention was given during the 1950s to differences in conditions and needs in the Third World until these conditions appeared to create obstacles to achieving high levels of industrial output. Leaders of multilateral and bilateral aid agencies strongly believed that the same processes of industrialization that brought economic growth to Europe would bring growth and modernization to developing nations. Development theorists were confident that the benefits of growth would eventually reach the vast majority of the poor through market mechanisms and trickle-down effects. Thus, the need to channel aid to the poor was obviated.

The industrialization policies prescribed by macroeconomic development theorists during the 1950s and 1960s sought rapid increases in gross national product (GNP). The only real debate was over the means by which they would be achieved. Some theorists argued that the most effective way of attaining high levels of economic growth was through heavy investment in capital-intensive industry or agriculture as a leading sector (Hirschman 1959). Others contended that a "big push" was needed in all sectors at the same time to increase output and demand for industrial production (Nurkse 1953). Both theories were modeled on processes of economic growth fueled by the industrial revolution in Western Europe and North America from the mid-1800s to the early 1900s. Developing nations were urged to seek large amounts of foreign capital, promote specialization in low-wage or raw-material-oriented industries, and apply capital-intensive technology to the production process. Orthodox economic theorists usually favored export production over import substitution. As industrial output grew it would generate more employment and higher incomes, which in turn would raise the level of demand for agricultural and industrial goods, increase savings, allow for expanded capital formation and generate new investment. Public expenditures, bolstered by foreign aid, would be concentrated in building physical infrastructure that would lower production costs and improve distribution. Nearly three-fourths of the loans made by the World Bank and the International Development Association from 1946 to 1963 were for physical infrastructure projects, especially for transportation, electrical power, ports and harbors, and industrial plants (Mason and Asher 1973).

One group of economists – leading sector theorists —insisted that the most effective means of generating and sustaining high levels of economic growth was by investing heavily in a single sector. Industry was usually considered the "engine of growth" for developing economies, but some economists argued that other sectors might be more appropriate for initiating investment. Some, such as Arthur Lewis (1954, 1955) and Theodore Schultz (1964), noted that agricultural investment would provide income for industrialization in rural nations; and still others, such as Currie (1966), insisted that growth in housing and consumer goods sectors would stimulate demand for industrial products and set off the spiral of investment.

But many economists, such as Albert Hirschman (1959), maintained that it mattered little which sector was chosen initially because heavy investment in any sector would generate increased demand and induced investment in all other sectors. Growth in the leading sector would spread and thereby raise the overall level of economic output. Hirschman argued that heavy public investments in either directly productive or social overhead activities would lower costs and, through complementarities in the economy, create increased demand and pressures for mobilization and investment of private capital. The ripple effect from this initial stimulation would generate growth throughout the economy. Thus, the objective of leading sector theorists was to create a set of continuous tensions in the economy: a sequence of events that "leads away from equilibrium is precisely the ideal pattern of development," Hirschman (1959: 66) insisted. Unbalanced growth would generate and enlist resources and abilities for development that had previously been "hidden, scattered or badly utilized." The mechanisms by which growth would spread were thought to be largely automatic once investment began. "If such a chain of unbalanced growth sequences could be set up," Hirschman (1959: 72) predicted, "the economic policy makers could just watch the proceedings from the sidelines."

Other economists argued for a different approach. They noted that in labor-surplus economies investment in one industry or set of industries could not generate sufficient employment, income, and demand to absorb output. Moreover, heavy investment in one industry precluded development of other sectors that provided inputs for industry. They argued that massive amounts of foreign aid should be used in combination with national resources to make a "big push" for development by investing simultaneously in all sectors. Balanced investment would create the internal complementarities that Hirschman and others assumed already existed in developing economies. It would allow each sector to supply the others without heavy reliance on imports. Natural resources could be used appropriately, employment would increase throughout the economy, and greater demand would be created for outputs in each sector. Agriculture and commerce would benefit as well as manufacturing. Moreover, expansion of industry would proceed at pace with improvements in labor skills and entrepreneurial experience. Investment in physical infrastructure, public utilities, production equipment, and plant would be balanced, but sizable enough in each sector to push the economy into a stage of rapid and sustained growth. Such a strategy was necessary, Nurkse (1953) insisted, because in the long run productive capacity, level of production and ability to use capital to increase output were all limited by the size of the market, which was extremely small in most poor countries.

The question of how massive poverty – a major factor limiting the size of markets in developing nations – would be alleviated, was rarely asked by development theorists or directly addressed in aid strategies. The problem of reducing the large gaps in income and wealth between rich and poor nations would be solved, Rosenstein-Rodan (1943) argued, by achieving "a more equal distribution of income between different areas of the world by raising incomes in depressed areas at a higher rate than in the rich areas." This would occur internationally through the same automatic mechanisms that Hirschman relied upon in national economies. Investment in any activity would set in motion complementary stimuli that would spread growth throughout the economic system, thereby eventually alleviating poverty.

As industrial production increased in developing countries new jobs would be created, demand for new products would be expanded, and through forward, backward and lateral linkages new investments would be made. This would create additional employment opportunities and raise overall levels of income. Some of the new income would be spent on basic needs for food, shelter, education, and health. Some would be taxed to provide public services. Some would be saved and reinvested. Increasing employment would not only draw larger numbers of people into the productive system, but the resulting demand for labor, goods, and services would spread from the major urban centers where large-scale industries were located, into smaller towns and rural areas. Increasing incomes would create higher demand for agricultural goods, and the application of new technology by farmers would make agriculture more productive and less labor-intensive. Surplus agricultural labor would be absorbed in the expanding industrial sector. As agricultural production increased, profits would be reinvested in more efficient technology, better seed varieties, irrigation, and other inputs that would generate even higher yields with less labor and land. Once the economy reached the "take-off" stage, more of the poor would begin to benefit and a growth cycle would generate higher levels of output. These forces would create incentives for diversification and allow more technologically advanced industries to succeed lower-wage and raw-materials-oriented industries (Rostow 1952).

Many economists believed, along with Kuznets (1966) who formalized the theory, that in the initial stages of growth the largest share of income would go to higher income groups, but that the income share of the poor would increase as growth continued. When growth was rapid enough to change the dualistic structure of the economy, a more equitable distribution of benefits would begin to eliminate destitution. Many economists argued that reallocation of investments to generate a wider distribution of income by channeling aid to poor groups in developing societies would slow the overall rate of economic growth and thus delay the time at which the poor's share of income would begin to rise on the "Kuznets curve."

RELIANCE ON LONG-RANGE, CENTRALIZED, NATIONAL PLANNING

Both the leading sector and big-push theories relied on comprehensive, long-range planning by national governments to formulate and implement development policies. Macroeconomic planning became the main instrument for achieving growth policies and a precondition for international aid to poor countries.

Macroeconomic planning became fashionable following the Second World War. Although pre-war experiments were not uncommon, they were limited and prompted mostly by the Soviet Union's apparent success in mobilizing investment resources through central planning and by European governments" attempts to rationalize and control colonial economies in the developing world. The spread of national planning in the late 1940s and early 1950s can be more directly attributed to the effectiveness of war mobilization planning in Europe and to the desire of governments attaining post-war political independence for rapid economic growth. But perhaps the greatest impetus to national planning was the insistence of international aid agencies that grants and loans be made in conformance with coherent plans for national development. Thus, serious attempts to plan and manage economies came in the wake of World Bank economic missions to developing countries in the 1950s. Creation of the Colombo Plan at about the same time provided additional incentives for national planning in India, Pakistan, Singapore, and Sarawak. Attempts in the early 1960s by United States aid officials to mandate national planning as a precondition for assistance stimulated central analysis in Korea, the Philippines, and Taiwan (Mason and Asher 1973, Waterston 1965).

The influence of international lending agencies and economic theorists was reflected clearly, for instance, in the objectives and procedures adopted by Asian planners. Although national planning evolved in different ways in different political systems, its goals and procedures nonetheless were strikingly similar. "The basic principle in the ideology of economic planning," Myrdal (1970: 175) noted in his extensive study of Southeast Asia, "is that the state shall take an active, indeed, the decisive role in the economy; by its own acts of investment and enterprise, and by its various controls – inducements and restrictions – over the private sector the state shall initiate, spur and steer economic development." Policies thus would be "rationally coordinated, and the coordination [made] explicit in an overall plan for a number of years ahead."

For most countries attempting to establish political and economic independence following the war, central planning offered not only an efficient tool for allocating scarce resources, but also a symbol of progress and self-control. By the mid-1960s Waterston (1965: 28) could observe with little exaggeration that "the national plan appears to have joined the national anthem and the national flag as a symbol of sovereignty and modernity." But beyond mere symbolism, Asians turned to central planning as a means of quickly achieving their economic and political aspirations. Most of the Asian countries" early plans reflected goals similar to those of Malaysia's first five-year plan. Drafted in 1954 with the assistance of the World Bank Economic Mission, it sought to stimulate industrialization, expand public facilities and infrastructure, and create employment for a growing labor force. Through national planning the government would rapidly accelerate public and private investment, especially in export industries, and increase the rates of capital formation and savings (Rudner 1975).

In Indonesia, which initiated its eight-year development plan in the late 1950s, central planning was an instrument of nation-building in the post-colonial period. President Sukarno designed a plan aimed at increasing consumption, self-sufficiency in food and clothing production, and basic infrastructure and utilities. Promoting self-sufficiency in basic commodities was motivated more by political than economic concerns, resulting from Sukarno's desire to disassociate Indonesia from its colonial past and to evoke self-sacrifice, political support, and national solidarity among a diverse people (Humphrey 1962). As the ideology of national planning spread, Asian governments assigned to it increasingly complex and diverse objectives, but the major goals remained those of accelerating investment and the growth of gross national product.

During the 1950s and 1960s Asian national planning took three basic forms: (1) top-down planning, through which a central planning agency formulated policies based on macroeconomic, quantitative models for the national economy; (2) bottom-up planning, through which the central planning agency compiled and reviewed the investment proposals of national ministries, local governments, and semi-public corporations and allocated resources to them on the basis of centrally determined economic priorities; and (3) mixed systems, which used a combination of top-down and bottom-up approaches.

Top-down planning usually began with designation of broad national development goals and targets. Macroeconomic analyses and econometric models sought to forecast long-range conditions and – based on predictions concerning operation of the economy and the influence of exogenous variables – to compare development objectives with forecasted conditions. Investment resources were then allocated to sectors with projected shortfalls. Macroeconomic studies, the formulation of alternative targets and strategies, and preliminary design of the national plan, were usually the responsibility of a central planning agency reporting to the prime minister or a council of ministers. The final plan, based on initial policy recommendations and tempered by other political and economic factors, sought not only to control public agencies" invest- ment decisions, but to guide their operating and budgeting decisions as well.

The South Korean plans, for instance, in addition to analyzing current economic trends, outlined the content and size of the government fiscal budget and recommended sectoral investment levels and incentives for private enterprise. They attempted to forecast major monetary trends, supplies of and demand for important commodities, foreign capital imports, price fluctuations, and significant private sector activities. In addition to establishing targets for public and private investment, the plans set foreign transaction, industrial, population, employment, and science and technology policy for government ministries and publicly controlled institutions (Korea, 1971). Thailand's development plans during the 1960s were based on econometric models consisting of five functional equations and five identities to project 15-year trends in gross domestic product, population, capital formation and savings, imports and exports. Major recommendations for resource allocation and investment policies were based on gaps between desired targets and forecasted conditions (UNDP 1972).

The "bottom-up" approach used in other countries began with submission of proposed projects by operating ministries, quasi-public corporations, and state, provincial, or local governments. In Pakistan, for instance, investment programs originated at the local and provincial levels. Local agencies identified projects that were reviewed and integrated into sectoral plans by provincial ministries, evaluated and processed by the provincial planning agencies, and transmitted to the National Planning Commission for consolidation into a national plan. In India, working groups similar to those organized at the national level were formed within state governments to identify, generate and initially formulate investment proposals for inclusion in the national plan (UNDP 1969).

Still other countries used combinations of bottom-up and top-down planning, or switched from one procedure to the other after evaluating initial results. Thailand used top-down procedures for earlier plans and bottom-up processes later, eventually merging elements of both. The National Economic Development Board (NEDB) in Thailand abandoned econometric modeling in the 1960s and was reorganized to coordinate development policies with operating and budgeting decisions. NEDB was formally charged not only with conducting studies of national socioeconomic trends and estimating resource availability, but also with reviewing the investment proposals of various ministries and public agencies. Its staff evaluated potential projects and integrated approved proposals into a national development budget that was submitted to the Council of Ministers for final approval (Changrien 1970).

The rationale for national planning in nearly all Asian countries was that since the public sector was the dominant force for development in capital-scarce countries, its allocation and investment decisions would have to be rationally, efficiently, and objectively planned in order to stimulate the economy. National planning would not only establish a coherent overall framework for public resource allocation, but would also establish guidelines for decision-making by public and quasi-public agencies concerned with national development, and provide criteria for evaluating private sector investment.

But Asian experience with national economic planning provided little evidence that it achieved either goal: economic growth in Asia during the 1950s and 1960s was sporadic and limited to a few countries, and the administration of national economic planning was plagued with severe problems, limiting its usefulness in guiding or controlling investment decisions.

THE LIMITS OF COMPREHENSIVE PLANNING

Theories of economic development and the strategies of international assistance organizations overlooked the fact that governments of most poor countries lacked the analytical and administrative capacity to formulate and implement the comprehensive, long-range national development plans required to achieve in a short time what had been attained gradually over a long period in the west. The ability of governments to coordinate their activities among a variety of ministries and agencies to implement a "big push" or leading sector strategy was weak. Few governments could manage public investments effectively, let alone guide or control those of the private sector. The experience in Asia is instruc- tive, for national planning and macroeconomic development policies did little to promote growth in nearly half of the Asian economies in the 1950s and 1960s, and the progress that was made in four of the high-growth economies – Hong Kong, Singapore, Taiwan and South Korea – had little to do with long-range national development plans (Rondinelli 1978).

The failure of national planning to achieve the goals of rapid economic growth and more rational and efficient investment decision-making in Asia can be attributed to two major factors: first, the limited administrative capacity of most Asian governments to implement highly centralized economic plans; and second, unrealistic assumptions and expectations concerning the power of national planning to guide and control national development.

Analysis of plan formulation and implementation in Asia during the 1950s and 1960s reveals a host of recurring and mutually reinforcing administrative problems. Among the most serious were: (1) the lack of strong political and administrative support for comprehensive plans; (2) deficiencies in the substantive content of plans that weakened their influence on resource allocation and investment decision-making; (3) the ineffectiveness of macro-planning methods and techniques; and (4) weaknesses in government administrative structure and procedures that limited their implementation capacity and their ability to control and evaluate the plan's results.

Weaknesses of Political and Administrative Support

Successful central planning requires political support – the willingness of high-level administrators and officials to make and carry out investment decisions that adhere to the national plan. Yet in most Asian governments strong political and administrative support for comprehensive planning was the exception rather than the rule. The shallowness of support was reflected in the opposition of ruling elites to plans suggesting fundamental changes in social and economic structure, the difficulty of planners in establishing their authority or in legitimizing their plans with administrative agencies, the inability to mobilize popular support for major development policies, and the frequent lack of communication between planners and administrators. Indeed, the most crucial decisions regarding social and economic development in Asia were often made outside the formal planning process.

The lack of strong political and administrative commitment appeared recurringly as a problem in evaluations of Asian planning. The "basic problem of development planning in Thailand," one official noted, "has been an absence of any keenly felt need for planning" during periods of satisfactory growth. Periods of economic stagnation produced demands for more immediate action than long-range planning could satisfy (Unakul 1969). Another analyst observed that "planning has never had a glorious day in the Philippines." Not only did it lack strong political support, but "planning as an attitude of mind, as an institution with all the sense of anticipation, cohesiveness and national discipline that it involves has not been socially accepted." Support was weak throughout the Philippine government structure. "There is no habit for it, no real experience in it," Valdepenas (1973: 265) argued, "and only a lot of attempts at escaping the consistencies and rigors it implies whenever efforts are made to apply it with some seriousness." In Sri Lanka, the political elite who were opposed to rapid social and economic change did little to back national plan priorities. "The absence of strong commitment or even a sense of continuity in development efforts," LaPorte (1970: 167) discovered, "has seriously affected public sector activities."

Describing two decades of national planning in Nepal, Rana (1974) noted that few national development decisions were influenced by the National Planning Commission. "The Planning Institution has undergone a series of chameleon-like changes ranging from full executive authority and involvement to a mere advisory role," he contended, and went on to note:

> Thus, despite its long history and the four plans it has produced, it is difficult to speak with confidence about its level of institutionalization. Projects whose feasibility the Commission has not analyzed continue to be taken up in the nation's program. The major policies of many sectors are often initiated and financed elsewhere.
>
> (Rana 1974: 660)

Cole and Lyman (1971) pointed out that early Korean plans did little to increase the government's effectiveness. The acceptance of national planning in Korea seems to have had little to do with increased commitment to comprehensive analysis. Rather, the succession to power of a military regime, optimism resulting from rapid economic growth, pressures from international lending agencies, the cohesiveness of Korean society, and the close identification of business interests with government development objectives made central controls more tolerable.

Administrators in Thailand not only withheld support, but often attempted to avoid or subvert national planning procedures, perceiving them not as instruments for promoting economic growth, but as ways "to advance the regime's political interests" (Nophaket 1973: 205). Thai planners who tried both top-down and bottom-up approaches were constantly frustrated in their attempts to integrate national planning with administrative decision-making. Indeed, the first plan entirely ignored private investment and dealt summarily with the microeconomic and financial factors of most concern to ministry officials. Ministries and quasi-public corporations often succeeded in avoiding the central planning agency's project review processes by financing investment proposals from operating budgets.

Deficiencies in the Content of Plans and Failure to Identify Investment Projects

In most countries the plans themselves lacked substance to guide decision-making and rarely identified specific projects. Most Asian plans stated objectives in vague and amorphous language calculated to gain widespread consensus without specifying implementation strategies, which might generate conflict and opposition. They often lacked cost estimates and resource allocation proposals, failed to disaggregate macroeconomic targets through intermediate sectoral or regional plans, and did not identify programs and projects for funding in annual or capital budgets. Thus, few plans provided useful guidance for mobilizing and allocating domestic resources or for distributing foreign aid and other external capital to high-priority programs.

Despite the claim that national planning would lead to decisions that were more systematically and comprehensively analyzed and that central review would provide rationally ordered priorities for the allocation of scarce resources, few national plans actually attained those goals. One evaluator of early Malaysian plans noted that:

> [They] were no more than aggregations of the expansion programs of separate governmental departments. The planning procedure was simple. Each government department was requested to submit its own claim for recurrent and development expenditure. The total of these claims would normally exceed the funds available and a committee, acting under certain unwritten rules and criteria, would reduce the sum total of the claims until it equalled that of the funds available.
>
> (Lim 1973: 90–1)

The decisions were made without formally announced priorities and "with no regard for internal consistency." When subsequent attempts were made to correct these deficiencies by adopting quantitative macroeconomic models, however, the plans failed to disaggregate, or to predict accurately, investment needs for specific sectors.

National plans were particularly weak in specifying investment priorities in the private sector, which proved to be the most dynamic part of many Asian economies in the post-war period. In Sri Lanka early plans paid "very little attention . . . to initiating, formulating and assisting investment in the private sector" (Karunatilake 1971: 270). There was virtually no relationship between private investment decisions and national plan recommendations in Nepal in the late 1960s. "None of the industries for which targets had been provided by the Third Plan fulfilled their targets," Rana (1974: 50) pointed out. "In fact, more than half of them were not even under production in 1970." Other industries succeeded beyond expectations, even though the plan set no targets for them. "In fact, success, where it was achieved, was either due to foreign aid negotiation totally independent of the plan, or was, in the case of stainless steel and nylon yarn, ad hoc decisions in complete contradiction of plan aims."

Evaluations also revealed the weaknesses of national planning in linking development recommendations to public budget decisions. Reviewing planning in Thailand during the 1960s, one official argued that "no consistency check was attempted for the plan nor were criteria laid down for selection of projects, including cost–benefit analysis" (Marzouk 1972: 431), and, according to another observer:

> Apart from the formation of sectoral programs, estimates were made for the production of major commodities at the end of the plan period. However, the final "target" of five percent growth of GNP was more a forecast than a target. Besides the government was not in a position to ensure that the entire program set out would actually be implemented. Besides, a considerable shortcoming in expenditure resulted. Nevertheless, the five percent "target" was achieved, even though it was largely due to the remarkable performance of the private sector, for which the plan had little to say.
>
> (Unakul 1969: 68)

Budget decisions in the Philippines rarely had any relationship to plan priorities and recommendations. Indeed, budget decision-making was often totally divorced from national planning. Budgets were worked out incrementally, through bargaining and negotiation on the basis of immediate political and administrative criteria. Although the Budget Commission was supposed to be guided by the plans of the National Economic Council (NEC), they were not always endorsed by the president. The Budget Commission instead prepared annual expenditure authorizations on the basis of needs submitted by operating departments. Appropriations were then either cut or increased during reviews by the president's staff and again during Congressional consideration, usually without reference to comprehensive development plans. Moreover, the Central Bank set monetary, credit and foreign exchange policies independently of both the NEC and other planning agencies (UNDP 1972). And even where attempts had been made to focus on specific investment implications, as did Indonesia's plans of the early 1960s, political goals often dominated and biased the decisions. Indonesia's initial plans, seeking to create national political solidarity, "grossly under-represented the financial costs and the economic sacrifices necessary" for implementation, even after control passed securely from the colonial to the national government

(Schulz 1972: 62). In Nepal, government officials complained of the "lack of relationship between the budget . . . and the Five-Year Plan" (Rana 1974: 51). Reviewing the effect of planning on budgeting in developing countries, Caiden and Wildavsky (1974: 251) concluded that "the annual budget rarely does what the plan intends." They found large discrepancies between planned investment and amounts actually budgeted. Singapore either exceeded or fell short of planned investments in each sector. Malaysian investments in infrastructure and utilities outstripped targeted allocations during the first three national plans, and investment in social services and agriculture, which received the highest priorities, failed to reach projected levels.

Ineffectiveness of Central Planning Methods and Techniques

The adoption of western rationalistic techniques of planning often rendered national economic plans inoperable. Evaluations of Asian planning suggest several problems. Models and techniques developed in western industrial countries were often indiscriminately applied in Asia. The macroeconomic models used in Malaysia's national development plans were found to be both inappropriate and inaccurate for forecasting national economic trends. Criticizing the use of an aggregate Harrod–Domar model, one Malaysian economist (Lim 1973: 121) pointed out that "even if the model were representative of the structural conditions of Malaysia, its usefulness would be reduced considerably by the shortage of reliable statistical data". The model's fundamental assumption, that the shortage of capital was a crucial bottleneck to growth, did not apply in Malaysia, and the data required to calculate the incremental capital–output ratio (ICOR), on which the model was based, had to be approximated using arbitrary – and what later proved to be inaccurate – assumptions.

A leading Thai planner (Unakul 1969: 74) noted the limitations of quantitative analysis under conditions found in most Asian countries, where "available data are neither comprehensive, reliable nor timely." One evaluator of Thailand's national plans for the 1960s claimed that not only were the macroeconomic models based on questionable and simplified economic assumptions, but that the World Bank Consultative Committee prepared the quantitative forecasts "basically by intuition." Marzouk (1972: 437) pointed out that "rough guesses had to be made to estimate GDP by expenditure type because of the paucity of data." And the director of the NEDB's Economic and Social Planning Division (Unakul 1969: 70) noted that private sector investment and consumption had to be considered as residuals by making the assumption that production of and demand for goods and services would remain in equilibrium through the end of the 1960s. That procedure, he contended, led planners "to assume away the central problem of planning, namely, how to adjust the resources and expenditures in both the public and private sectors to meet the objectives of growth and stability."

Inadequate Administrative Capacity for Central Planning and Control

Finally, coordination of administrative decisions concerning the allocation and investment of national resources must be an essential element of national planning if it is to achieve its targets and goals. Unless administrative agencies, quasi-public corporations, and private investors are willing to cooperate with the central planning agency in carrying out planned activities and to coordinate with each other in implementing programs and projects, central planning is meaningless. Yet, in reality, plan implementation in Asia was always severely limited by inadequate administrative structures and coordinative procedures for guiding or controlling investment decisions.

In part, the problem was due to the substantive deficiencies in the content of plans – their failure to specify investment needs and to anticipate administrative requirements for implementation. Rana (1974: 23) complained that the effectiveness of planning in Nepal was severely limited by the failure of planners to analyze the administrative and political implications of their proposals. Concern with macroeconomic analysis and quantitative forecasting, along with the application of western development theories, blinded them to the weaknesses of indigenous administrative capability. Arguing that the low level of managerial capacity was the most crucial bottleneck to development in Nepal, Rana concluded that "management, organization and enterprise are matters not precisely quantifiable and it is perhaps not surprising that westerners, and westerners highly specialized in economics, looking at our countries should have tended to take these factors for granted."

But it was precisely these organizational and administrative factors that appeared repeatedly in evaluations as critical constraints on the implementation of national plans. In Sri Lanka economic achievements always fell behind planned targets; decisions regularly departed from government guidelines. "More emphasis seems to have been devoted to the preparation of elaborate paper documents," Karuntilake (1971: 263) complained, "rather than to the project content of the plan and how best it could be implemented to produce quick results." And even after 1965, when the government explicitly admitted that national planning had little influence on essential development decisions and disaggregated planning by sector, progress was still limited by administrative constraints. Neither the Department of National Planning nor the ministries had the management procedures or technically qualified manpower needed to prepare and implement sectoral programs. After a decade of comprehensive planning, they discovered that even within the national planning agency very few officials had the experience necessary to determine balance-of-payments effects or GDP implications of sectoral plans.

To the extent that national planning required cooperation among ministries and with the central planning agencies, or depended on the ability of a central agency to coordinate, monitor, and evaluate resource allocations and investments, it was rarely successful in Asia. A Pakistani planning official (Hussain 1973: 458) reported that a major obstacle to executing that country's bottom-up process "is the total absence of coordination among various departmental agencies during implementation of projects." In Thailand, the Budget Bureau made fiscal and financial policies using its own criteria, leaving the national planning agency uninformed of its intentions. Despite formal requirements that it cooperate, the Budget Bureau did not usually provide revenue and expenditure information necessary for coordination between planners and budget officials (Unakul 1969: 72). Difficulties in implementing national plans also arose from the inability of central government agencies to evaluate, monitor, and control the decisions of administrative agencies. Even when planning and administrative structures were reorganized for that purpose, the results were disappointing. Thailand's requirement, that each operating ministry and government agency submit quarterly evaluation and review reports, quickly inundated planning and budgeting officials with more information than they could possibly analyze or that the Inspector General could investigate – even if the planners were able to find discrepancies between plans and operations. Indeed, Changrien

(1970: 24) noted that administrative reorganization of monitoring functions did little to improve coordination even among the review agencies:

> There is hardly any coordination between the Bureau of Planning and Research, the Budget Bureau and NEDB in this supervision and coordination work although all of them are in the Prime Minister's Office. The net result of this lack of coordination and ineffectiveness of the reporting system is that the Government Performance Information Center has almost ceased to be operative and the back-log of reports . . . in the Bureau of Planning and Research continues to mount. Supervision cannot be exercised over the government operations effectively and coordination is often left to chance.

In Malaysia, the government established "operations rooms" in an effort to evaluate and supervise plan implementation. They were effective only as long as the Prime Minister placed political pressure on the bureaucracy to report progress and coordinate activities. But the system itself was quickly bureaucratized. As political pressure from the Prime Minister eased, the system faltered (Esman 1972).

In brief, the failure of comprehensive national planning, either to achieve a clear record of success in promoting economic growth or to guide and control resource allocation and investment decisions, prompted the search in the 1960s for more effective policy-making instruments. The dismal record of national planning in the Third World led to a fundamental reappraisal of both planning procedures and macroeconomic development theories.

OBSTACLES AND BOTTLENECKS TO DEVELOPMENT

By the early 1960s it became increasingly apparent that in most developing nations a strategy of rapid growth through capital intensive industrialization was not working. Growth occurred in some Third World nations during the 1950s and early 1960s, but at rates well below those sought in national development plans. Studies found that foreign aid had little direct impact on increasing the levels of GNP in less developed countries. For instance, Griffin and Enos (1970) discovered that the correlation between foreign aid and increases in GNP during the 1950s was weak or insignificant for poor countries in Africa and Asia. Aid and growth were negatively correlated in Latin American countries, where "the greater inflow of capital from abroad, the lower the rate of growth of the receiving country." Governments had difficulty obtaining large amounts of foreign capital to finance ambitious industrialization plans and had little success in mobilizing sufficient savings internally to achieve high rates of capital formation.

Some theorists began questioning the model underlying capital-intensive industrialization theories. Dudley Seers (1965), among others, pointed out that the prevailing conditions that made rapid and sustained industrialization possible in western societies constituted a "special case," and that economic development policies applied successfully under those conditions could not simply be transferred or replicated in developing countries. He noted that in Europe and North America factors of production had been abundant during their periods of industrialization; labor was educated, skilled, mobile, and could easily be organized for productive purposes. Land was also abundant, arable, and widely held in private ownership. Capital was readily available; most sectors of the economy were heavily capitalized; entrepreneurship was well established; and governments provided inducements for entrepreneurial expansion. The structure of the economy was diversified and dominated by a competitive manufacturing sector that had evolved from numerous cottage and artisan enterprises. Agriculture was largely commercialized at the time of the industrial revolution in these countries and an extensive marketing network had evolved providing farmers with accessible and competitive outlets for their products. Public revenue collection and allocation procedures were firmly established. Savings could be mobilized by an efficient banking system in countries where the level of national investment was already high. Exports were diversified in products for which there was also an internal market. Income distribution was relatively equitable and a comparatively small percentage of the population remained destitute. Population growth was below 2 percent a year and a large percentage of the people lived in urban areas (Hoselitz 1964).

But in developing countries the conditions were far different. Low levels of education and skills within developing societies limited the availability of productive labor. Poor countries lacked large numbers of experienced managers. Low levels of agricultural production and massive poverty in rural areas limited the expansion of internal demand, leaving industries in developing nations at the mercy of export markets in which they were often uncompetitive or at a severe disadvantage. Moreover, poor countries had to pay substantially higher prices for their imports of equipment, technical know-how, intermediate goods and finished products than they received for their exports, leading to serious deficits in their balance of payments. High population growth rates offset advances in output or income, leaving much of the population no better off economically at the end of the 1950s than a decade earlier (Streeten 1972).

Market mechanisms that were supposed to act as channels for the spread of growth impulses and the filtering down of benefits worked imperfectly in many developing countries. Instead of growth spreading throughout developing economies, resources were often drained from rural hinterlands, through what Myrdal (1957) called "backwash effects," to support industries located in metropolitan centers. Political instability, low levels of administrative capacity, pervasive corruption among politicians and bureaucratic elites, and an unwillingness of political leaders to share power, or to enforce laws that would maintain order with justice, led to the creation of "soft states" in which governments were unable to organize society for developmental purposes (Myrdal 1970). Productive assets such as land were generally owned or controlled by a small, privileged elite who opposed reforms that might lead to a greater distribution of income and wealth. Their profits were often invested in the largest urban centers or outside the country.

Under these conditions entrepreneurship could not easily be promoted and the low levels of income received by the vast majority of people continued to inhibit the expansion of internal demand. With large subsistence populations, it was extremely difficult for governments in developing countries to mobilize savings or generate revenue through direct taxes. Thus, the level of public investment depended on revenue raised through indirect taxes, export earnings, foreign aid, and external borrowing. As a result, investment remained a small percentage of gross domestic product.

The spread or trickle-down effects of growth were constrained by weak market and trade linkages between major industrial centers and rural areas. Many developing nations had "primate city" spatial structures: the bulk of modern economic activities, social services, infrastructure, and facilities had been concentrated in the capital city or in a single large metropolitan area which dominated the spatial system and economy of the nation (Rondinelli and Ruddle 1978). Few secondary cities could emerge, and large disparities in income and wealth arose between the primary city and the rest of the nation (Rondinelli 1983b). Disparities between the largest city and the rural hinterlands pulled large numbers of younger, more ambitious, and better educated rural people into the metropolitan center and they were followed by less educated or unskilled relatives and friends who often could not find jobs in the city or had access only to the lowest paying employment. Slums and squatter settlements grew quickly in the cities.

ECONOMIC GROWTH THROUGH SOCIAL CHANGE: OVERCOMING OBSTACLES TO DEVELOPMENT

Not only did capital-intensive industrialization strategies of the 1950s not produce rapid and sustained growth in developing countries, but in many it created "dual economies" and reinforced a cycle of poverty that became more difficult to break. The foreign aid strategies of the 1950s strengthened the forces perpetuating poverty in developing nations, and by the end of the decade came under attack from both liberal and conservative economists in the United States. Milton Friedman (1958), for instance, supported the concept of foreign aid but questioned the three major assumptions underlying capital-intensive industrialization policies. He challenged the propositions that availability of capital was the key to economic development, that underdeveloped countries were unable to mobilize capital internally, and that centralized, comprehensive macroeconomic planning was a prerequisite to development. "All three propositions are at best misleading half-truths," Friedman argued. He noted that developing nations had been mobilizing capital and other resources for high priority investments for centuries. But, he insisted, other conditions were more important for promoting economic development and these had to be created before capital could be used effectively. Moreover, Friedman argued that the prescriptions for macroeconomic development planning were unlikely to be useful or appropriate in developing nations. He maintained that:

> Such a centralized program is likely to be a hindrance, not a help . . . Economic development is a process of changing old ways of doing things, of venturing into the unknown. It requires a maximum of flexibility, of possibility for experimentation. No one can predict in advance what will turn out to be the most effective use of a nation's productive resources. Yet the essence of a program of economic development is that it introduces rigidity and inflexibility.
>
> (Friedman 1958: 256)

Others insisted that bilateral foreign assistance tied to procurement in the donor country prevented governments from using the funds for high priority needs and for non-industrial purposes. The conditions under which aid was given often precluded its use for investments other than show-piece projects and high technology production. Ultimately these capital transfers discouraged indigenous savings and internal capital formation in developing nations. The benefits of aid often went to a small elite in power who were unwilling to undertake social and economic reforms or to initiate programs that would break the bottlenecks to economic expansion and diversification. Griffin and Enos (1970: 325), reflecting on similar conclusions arrived at by World Bank officials at the end of the 1950s, claimed that "foreign aid tends to strengthen the status quo, it enables those in power to evade and avoid fundamental reforms; it does little more than patch plaster on the deteriorating social edifice."

International assistance agencies began to target their technical and financial assistance during the 1960s more carefully on specific problems and conditions in developing countries that were thought to be obstacles or bottlenecks to industrial expansion and economic growth. An underlying assumption of these new approaches, which had been evolving over a number of years, was that social changes had to precede economic expansion because adverse conditions in developing nations were obstacles to growth. More attention had to be given to programs for asset redistribution, institution-building, population and family planning, labor-intensive and small-scale industrialization, agricultural expansion and human resources development. Insistence on the formulation of comprehensive long-range master plans gave way to sectoral programming and increased attention to project formulation and design. Again, both liberal and conservative economists hailed flexibility as an essential element of development planning. Friedman argued for development policies that would create an economic and political environment more conducive to widespread participation in economic activities:

> What is required in the underdeveloped countries is the release of the energies of millions of able, active, and vigorous people who have been chained by ignorance, custom, and tradition. Such people exist in every underdeveloped country. If it seems otherwise, it is because we tend to seek them in our own image in "big business" on the Western model rather than in the villages and farms

> and in the shops and bazaars that line the streets of the crowded cities in many a poor country. These people require only a favorable environment to transform the face of their countries.
> (Friedman 1958: 256)

Others insisted that development programs would have to be strongly guided by national governments and international assistance agencies through sectoral planning and investment programming. Internal resources and foreign aid would have to be channeled into sectors where bottlenecks were greatest and into programs that would generate fundamental social changes.

The Shift to Sectoral Development

Development theories that evolved during the 1960s began to focus on the search for the key sectors into which national resources and international assistance should be channeled. Some economists, such as Schultz (1964), saw agricultural development as the new "engine of growth" in poor countries. Johnston and Mellor (1961) insisted that agricultural sector development would have a number of beneficial effects. It would increase food supplies needed to meet the nutritional requirements of the population in poor countries and maintain lower food prices as demand increased. Expansion of agricultural exports would provide higher incomes to farmers and peasants and increase foreign exchange earnings. Higher agricultural productivity would free surplus labor for industrial employment and generate capital that could be invested in manufacturing and industrial sectors to absorb the surplus labor. Increased income for rural people would create larger demand for both agricultural and industrial products.

But the semi-targeted aid strategies of the 1960s were based on more than a leading-sector investment theory. It was widely recognized that the linkages among sectors had to be created in developing nations and that aid had to be channeled into a variety of supporting activities as well as into sectoral production. Thus, aid agencies provided technical and financial assistance for research into new high-yielding seed varieties, irrigation system construction, improvement of agricultural training and extension programs, the creation of marketing systems, the organization of cooperatives and farmers" associations, and the initiation of agricultural credit schemes.

It also became clear during the 1960s that social, economic, and political problems were inextricably related. The existence of some conditions had to precede the creation of others and among the most important was the redistribution of productive assets. Some development theorists argued that breaking up the monopoly of land ownership was the most effective way of redistributing assets to the poor and making income distribution more equitable. The redistribution of lands owned by plantations and family estates to cultivators and peasants would reduce rents and increase income in rural areas while putting more land into cultivation (Warriner 1964). The US aid program not only funded many of the land reform programs in developing countries, but also sent technical experts to help carry them out. United Nations agencies urged governments to adopt land redistribution policies.

The low levels of administrative capacity in governments of developing nations also prompted international aid agencies to provide technical assistance in public administration. Public administration specialists sought to reorganize bureaucracies, establish civil service systems based on merit and skill, improve personnel administration, establish public enterprises, and reform the budgeting and investment allocation procedures in Third World countries. They emphasized "institution-building" as a means of modernizing governments and of expanding their capacities to carry out development activities more effectively.

High rates of population growth were also seen as fundamental bottlenecks to economic and social progress in developing nations during the 1960s, and thus large amounts of aid went to private and voluntary organizations promoting population control and family planning. Aid was also channeled into human resources development. Assistance organizations worked with governments to establish formal, informal, and vocational education systems in developing countries. Harbison (1962: 3), among others, argued that human resources development "is concerned with the two-fold objective of building skills and providing productive employment for unutilized and underutilized manpower." Aid programs would be used to increase the supply of professionals and scientists in developing countries, prepare young people to take up technical positions for which para-professional training was needed, and relieve the severe shortages of managerial and administrative personnel in the public and private sectors. Expanding the numbers of trained teachers, the shortage of which Harbison called the "master bottleneck" that constrained the entire process of human development in developing countries, received higher priority. Educational systems would also have to be reorganized to train people for occupational roles as craftsmen, clerical workers, and entrepreneurs. Large amounts of aid were also provided to send talented government officials and managers to the United States, Britain, and some European countries for graduate and professional degrees and for short-term professional and technical training. School systems in poor countries were reorganized to reflect American, British, or French standards and educational objectives.

The International Labor Office (ILO), concluding that unemployment had become "chronic and intractable" in most developing nations during the late 1950s and early 1960s, focused its attention on promoting labor-intensive, employment-generating industrialization, agro-industries, small-scale manufacturing, and informal-sector enterprises that would absorb labor surpluses in the Third World.

Sectoral Planning and Systematic Project Design

The shift from untargeted to semi-targeted aid was reflected not only in the sectoral development policies adopted by international assistance organizations during the 1960s, but also in their planning and implementation procedures. Both bilateral and multinational assistance agencies abandoned requirements for national, long-range, macroeconomic development plans and turned instead to sectoral planning and investment programming. They also gave greater emphasis to project preparation and design. The shifts came as the result of the growing acceptance of two basic assumptions in international assistance agencies that were made explicit by World Bank official Albert Waterston (1965). The first was that external assistance would have little impact unless it was more precisely tailored to the needs of developing countries and aimed more effectively at overcoming specific problems or obstacles to development through well conceived and efficiently organized development projects. National development plans had failed to disaggregate goals and objectives into programs of investment and to guide the selection of projects. At the same time, officials of international aid agencies assumed that governments in developing countries did not have the administrative capacity to use foreign aid effectively. Waterston (1971: 240) maintained that: "technical ministries, departments and aid agencies in most countries do not have the staffs qualified to (a) identify, evaluate and prepare good projects, (b) fix project priorities in accordance with well devised time tables, and (c) operate completed projects efficiently." But officials in international agencies also believed that individual projects and programs would not lead to coherent and well-focused development strategies and that aid had to be channeled into specific sectors in a highly coordinated fashion. Thus, as the World Bank's annual report for 1967 pointed out:

> an approach being increasingly employed in the developing countries is to undertake an analysis of a particular sector of an economy with a view to preparing a coordinated investment program for that sector and to selecting priority projects within it.
> (World Bank 1967: 16)

Aid strategies were to be formulated and implemented during the 1960s through two seemingly contradictory approaches. On one hand the World Bank and USAID experimented with program and sectoral lending arrangements that granted developing nations much more flexibility in using aid in priority sectors. These funding agreements did not tie aid to conventional projects. On the other hand all international assistance organizations adopted more complex and rigid requirements and procedures for identifying, preparing, appraising and implementing loans and grants that were made through projects.

The US Agency for International Development (USAID) began making sectoral and program loans in the early 1960s to a limited number of countries, mostly in Latin America, to finance imports for coordinated investment programs. Program loans were based on self-help agreements between USAID and the recipient government, in which the borrower pledged to: make detailed sectoral analyses; establish specific and precise goals for development of the sector; and outline proposals for policies, programs, and projects to be supported with foreign aid. The Latin America Bureau in USAID made the most extensive use of sectoral loans because officials were dissatisfied with the macroeconomic approaches to development assistance employed during the 1950s and with the rigidities of the project format. Latin America Bureau officials insisted that standard prescriptions for development and inflexible procedures for aid administration would inhibit social and economic change in developing nations (USAID 1972b). Between 1967 and 1972 USAID authorized twenty-one sector loans totaling more than $438 million, which were used primarily for agricultural development, land reform, employment generation, education, and municipal and urban development. Income redistribution was an explicit objective of many of the sectoral loans (USAID 1972c).

At the same time, international assistance organizations were adopting more complex and systematic procedures for planning and implementing development projects. Attempting to overcome deficiencies in planning and administrative capacity in less developed nations, aid agencies took a more active and direct role in project identification, preparation, design, and analysis. They formulated a complex set of procedures for testing preinvestment feasibility. To varying degrees the World Bank, USAID and the UN specialized agencies insisted on quantitative justifications of project proposals, including extensive market and needs analyses, technical feasibility studies, economic cost–benefit calculations or precise estimates of internal rate of return, and studies of administrative and managerial capacity to implement proposed projects. Preparation and selection guidelines were designed to ensure that proposals were compatible with lending institution policies and priorities (Rondinelli 1976a).

World Bank officials insisted, for instance, that effective implementation of development projects was influenced strongly by detailed preparation and appraisal procedures. "Careful project preparation in advance of expenditure is," Gittinger (1972: 1) insisted, "if not absolutely essential, at least the best available means to ensure efficient economic use of capital funds and to increase the chances of on-schedule implementation." Bank officials began investing substantial amounts of money in detailed feasibility analyses and project appraisals, insisting on thorough exploration of the creditworthiness of borrowing countries and of the financial feasibility of project proposals. They believed that the more elaborate and detailed the feasibility and appraisal analyses, the greater the probability that the projects would be implemented successfully.

But others argued that preparatory analyses were not sufficient to ensure successful implementation. More systematic and detailed management procedures were needed after a project was approved. Thus, a number of techniques were adopted from the fields of management science, operations research, and corporate planning that had been used by private firms and defense agencies in the

United States and Britain to guide the management of development projects. Techniques that applied mathematical logic to organizational performance; cost-effectiveness analyses; precise time-phasing of project activities; and control of operations through the use of computers and complex reporting requirements were all prescribed by management consultants within international assistance agencies. Linear programming, CPM-PERT scheduling and control techniques and mathematical models for predicting economic and social behavior in developing countries all became fashionable in managing development projects.

Re-evaluation of Aid Impacts and Procedures

Although the new international development policies and aid strategies were more successful in disaggregating and focusing aid on critical problems, the overall results in developing nations had been disappointing. On the positive side, progress was made in most countries in lowering the rate of population growth; a few countries such as Taiwan, Malaysia, Brazil and South Korea were achieving high rates of economic growth; agricultural production increased in many developing nations; and health conditions improved for higher- and middle-income groups in the developing world. Yet land reform failed in more nations than it succeeded; population growth still outpaced agricultural and industrial production in most poor countries; and high levels of illiteracy, poverty, child death, and malnutrition were common in Third World nations. Evaluations of the UN's First Development Decade found that progress had been slow and halting during the 1960s, and that the poor in most developing nations were no better off at the end of the decade than they had been at the beginning. Few nations had attained the levels of economic growth projected in their national development plans. A study undertaken for the World Bank (Pearson 1969) (The Pearson Commission) found that despite overall increases in economic growth in developing nations during the 1960s, in 1967 nearly 70 percent of the population (outside of mainland China) lived in countries where per capita income grew at less than 2 percent a year and where population growth rates generally exceeded increases in national output.

By the end of the 1960s international evaluation commissions became increasingly critical of the ability of assistance organizations to deliver aid effectively and to break the bottlenecks and overcome the obstacles to development. The Pearson Commission (1969) reviewed the entire aid system for the World Bank and found that if technical and financial assistance was to become more effective it would have to be disassociated from political and military objectives of the aid-givers, untied from procurement requirements, and channeled more flexibly through multilateral organizations. The Commission saw the increasingly rigid and formal bureaucratic procedures surrounding international assistance as a hindrance to its effective use in developing nations. It noted that developing nations often lacked adequate policies for using aid, "leading to duplication of requests and confusion among government ministries interested in furthering their own interests" (Pearson 1969: 169).

An examination of the United Nations" system of technical assistance by the Jackson (1969) Committee came to similar conclusions. Among the Jackson Committee's most severe criticisms was that foreign aid was not tailored to the needs of developing countries. Western practices and institutions – or solutions to problems in one developing nation – were often transferred to Third World countries without modification. "Instead of measuring and cutting the cloth on the spot in accordance with individual circumstances and wants," the Committee (p. 171) claimed, "a ready-made garment is produced and forced to fit afterwards."

But one of the largest stumbling blocks to the use of aid in promoting national development, the Pearson (1969: 29) Commission pointed out, was the failure of governments in most developing nations to distribute income and productive assets more equitably. "They have only recently begun to recognize that measures to make income distribution more equitable not only serve a social objective," the Commission argued, "but also are necessary for a sustained development effort."

3

DEVELOPMENT POLICIES AS SOCIAL EXPERIMENTS

From growth-with-equity to structural adjustment

Assessments of the performance of developing economies undertaken in the early 1970s reinforced emerging arguments that aid must be channeled to the poorest groups in developing societies if they were to benefit from and contribute to economic development.

In a review for the World Bank of the trends in economic development from 1950 to 1975, Morawetz (1977) found that growth was very unevenly distributed among developing nations and that statistics indicating overall progress in Third World countries often masked serious disparities. Although the average annual growth rate of GNP for eighty developing nations had been a respectable 3.4 percent – a rate that exceeded the growth rates of GNP in the industrialized countries (3.2 percent a year) – performance in individual developing nations varied widely. The developing countries" average had been raised by high rates of growth in oil-rich Middle Eastern countries (5.2 percent) and by rapid industrialization in a few of the East Asian economies (3.9 percent). But African and Latin American nations were able to increase their GNP by an average of only 2.5 percent a year and Southeast Asian economies grew at an average rate of only 1.7 percent a year. Many of the poorest countries were completely bypassed by economic progress. Honduras, Chile, Ghana, and Bolivia, for instance, hardly increased their economic output at all (less than 1 percent a year in the 1960s and early 1970s), and others such as Bangladesh, Rwanda, and Upper Volta registered negative growth rates.

Thus, statistics that showed substantial progress in economic development blurred the fact that disparities between rich and poor countries had actually increased over 25 years. Although growth rates remained stable for the eighty countries analyzed by Morawetz (1977), absolute disparities between industrialized and developing nations in per capita GNP increased threefold. In 1950, average GNP per capita in industrialized nations was $2000 greater than that in developing nations. By 1975 the differences more than doubled to nearly $5000. Even the economies that were growing the fastest during the late 1960s and early 1970s, such as Korea and Taiwan, were unable to reduce the gap substantially. Differences in per capita GNP between western countries and these fast-growing Asian economies doubled during the quarter century since 1950.

By the early 1970s there remained at least thirty-seven countries with a combined population of over 1.2 billion that had per capita GNP levels of less than $300 a year, and another fifteen countries with 270 million people with per capita GNP of less than $500. Nearly three-fourths of the population in less developed nations lived in relative or absolute poverty with income equivalents of less than $75 and $50 a year, respectively. Less than half of the population in the thirty-seven least developed nations were literate. Average annual population increases in these countries were more than double those in developed countries. Life expectancy averaged less than 50 years. Child death rates were eighteen times higher on average than in industrialized countries (World Bank 1979).

It became increasingly apparent during the late 1960s and early 1970s that the numbers of people living in dire poverty in developing nations was increasing rather than diminishing. The conditions of many living at or near subsistence levels had been progressively worsening. Development theorists and aid officials acknowledged that high rates of economic growth alone – with all of the inherent constraints on spread and trickle-down effects – would not ameliorate poverty in most developing countries. The conditions of the poor grew worse rather than better in all but a few of the countries that had achieved high rates of economic growth. Studies done in the late 1960s and early 1970s indicated that the spread effects from sectoral development policies were as sluggish as those produced by macroeconomic strategies. In a study of forty-three developing nations, Adelman and Morris (1973) found that, in all but the richest and the poorest countries, income distributions became more skewed as their economies grew. In many countries the highest income groups (the top 20 percent of the population) received more than 65 percent of the nation's income and the lowest 40 percent of the population received as little as 10 percent. They found a good deal of evidence that questioned the validity of the Kuznets curve. Adelman and Morris (1967) found that in many developing nations the richest groups had indeed benefited from initial economic growth. But even after the economy broke out of extreme dualism, the benefits went mainly to middle-income groups; little or nothing trickled down to the poor. The position of the poorest 40 percent of the population worsened both relatively and absolutely.

The accumulating evidence suggested that it was necessary to expand the participation of people living at or near subsistence levels in productive economic activities in order to stimulate equitable and sustained economic growth. Historically, the expansion of economic participation in both industrial and developing countries came through what Adelman called a "human-resource-intensive development strategy." In a study for the World Bank, Chenery (1974) found that the vast majority of the poor had been excluded from the benefits of economic growth "by a number of specific disabilities that can be summed up as the lack of physical and human capital and lack of access." These disabilities could only be overcome, he noted, by designing policies that both promoted equitable distribution of income and took account of the specific social and economic characteristics of the poor.

Adelman argued that few, if any, single-sector development policies alone would have a real impact on reaching the poor majority. Those developing countries that had achieved growth with relative equity had several common characteristics. They had political leaders who were strongly committed to development based on broad participation in economic activities. They received massive amounts of foreign assistance that allowed them to tackle a wide variety of social and economic problems simultaneously. They successfully carried out strong programs of asset redistribution prior to the period of rapid economic growth, thereby expanding the share of the population who benefited from development. Most importantly, they placed a great deal of emphasis on human resource development in the pre-growth period. From these lessons, Adelman (1975) and others concluded that "asset redistribution and the redistribution of opportunities for access to asset accumulation are a necessary first step for the initiation of equitable growth." They also argued, however, that growth and equitable distribution policies could not be separated. In countries where sluggish growth followed asset redistribution, the value of redistributed assets declined leaving the poor no better off, while the richest 20 percent of the population captured windfall profits (Adelman 1975: 70).

NEW DIRECTIONS IN DEVELOPMENT ASSISTANCE: GROWTH-WITH-EQUITY STRATEGIES

The changes that came about in international assistance strategy during the early 1970s evolved from a confluence of forces. For nearly a decade the purposes of and approaches to development assistance had been re-evaluated. A steady process of learning by scholars, practitioners, and policy-makers in both rich and poor countries brought many conventional assumptions of development theory into question, as had the frequent attacks on mainstream economic theories by radical and socialist thinkers. Although socialism and radical Marxism, which rejected mainstream economic development strategies, influenced development theory quite significantly, international assistance organizations did not embrace the more radical approaches embodied in dependency theory, liberation economics, and revolutionary political ideology. The rationale for "targeting" benefits to constituent groups of the poor, however, reflected the influence of social scientists who were more concerned with the nature, direction, and implications of social change than simply with the level or pace of economic growth, and of many economists who were critical of the conventional approaches used to assist developing countries during the 1950s and 1960s.

An earlier belief prevalent among development economists that economic growth and equitable distribution of income were conflicting goals was largely displaced during the 1970s by evidence that deliberate efforts to distribute income and wealth more equitably in countries such as Taiwan, South Korea, and Malaysia did not impair high levels of economic growth. In fact, those policies created a broader base of participation in economic activities that reinforced and accelerated growth. It was also more widely recognized that "automatic" mechanisms rarely produced the expected spread and filtering-down effects in poor countries and that policies had to be deliberately designed to incorporate peripheral areas and marginal groups into the economy if poverty was to be reduced.

Moreover, political leaders and scholars from developing nations began asserting their own interests and views in international forums. The steadily growing realization that problems in developing nations differed drastically from those faced by industrial countries brought about a fundamental rethinking of development policy that led to new and more pragmatic approaches to international assistance in the early 1970s that were clearly reflected in the US Foreign Assistance Act passed by Congress in 1973.

The Foreign Assistance Act of 1973, as noted in Chapter 1, shifted the focus of American aid from a program that had attempted to build economies in developing countries in the western image to one that would assist developing nations in following their own paths to development. The new aid program would give less emphasis to maximizing growth through macroeconomic policies and pursue what the House of Representatives" Foreign Affairs Committee called a "people-oriented, problem-solving form of assistance." It put aside the capital transfer approach in favor of using a wide variety of technical and financial instruments for solving social and economic problems. The previous emphasis in aid strategy on maximizing economic growth was tempered by a concern for dealing with those basic social problems that kept the vast majority of the population in Third World countries in poverty. For the first time international assistance organizations were to take into account the characteristics and needs of the poor as the primary beneficiaries of aid programs.

The arguments made for these new directions in development assistance were based on familiar humanitarian and political motives. The Foreign Affairs Committee of the US Congress pointed out that the economic, political and military security of the United States depended in part on the progress of nations in which large numbers of people lived in poverty and deprivation, and that aid was a means of influencing governments in developing countries to follow policies that were consistent with American foreign policy. The humanitarian appeal also continued to be strong. The House Foreign Affairs Committee noted in its report on the Foreign Assistance Act of 1973 that "we Americans annually spend six times what we allot for foreign assistance on cigarettes and other tobacco products. We spend three times as much for toys and sports supplies; three times as much for toilet articles and perfumes. Surely, our consumption-oriented society can spare something for fellow human beings who have virtually nothing" (*US Code Cong. Admin. News* 1973: 2811).

Above all, the changing strategies reflected a somewhat more sophisticated understanding of the dynamics of development, the constraints on economic and social change in developing nations, and the political and social forces that were influencing the processes of change in Third World countries. Although these strategies identified the poor majority as the primary beneficiaries of aid, the new policies were not conceived of, or intended to be, welfare programs. In its report to the Congress, the Foreign Affairs Committee argued that "we are learning that if the poorest majority can participate in development by having productive work and access to basic education, health care and adequate diets, then increased economic growth and social justice can go hand in hand." The president of the Overseas Development Council (Grant 1973: 65) insisted that "greater equality of opportunity to participate, rather than more aid of the welfare variety, is the most urgent need of the poor within countries (and of the low income states within the community of nations)." He noted that equity can be "more efficient than inequity and 'trickle down' in advancing growth in both rich and poor countries."

The new development assistance policies also recognized that economic growth and equitable distribution of income were not incompatible. An analyst for the Foreign Affairs Committee of the US Congress, examining the ILO's call for an employment-generating, basic needs, income-redistribution approach to development, noted the differences between the development theories emerging during the early 1970s and conventional thinking by pointing out that the debate was not "between advocates of growth and advocates of no growth or slow growth; it is between advocates of maximum growth in GNP regardless of how it is achieved and advocates of a growth path which puts to productive use the now underutilized labor of the poor" (Paolillo 1975: 5). World Bank President Robert McNamara, in his 1973 speech to the board of governors in Nairobi, justified the bank's new emphasis on reaching the poorest groups in terms of their productive potential and the contributions that the poor could make to national development.

APPROACHES TO AID TARGETING

The international development assistance strategies that emerged during the early 1970s began to focus technical and financial assistance for the first time on a specific segment of the population in Third World nations. The "poor majority," or the poorest 40 percent of the population, or the "marginal groups" in developing societies, as they were variously referred to by international assistance organizations, were to be the target groups of the new assistance policies. International assistance agencies initially faced the problem of identifying and defining these beneficiaries and then of finding means of channeling aid to them. Three basic targeting strategies emerged during the 1970s: (1) aid was channeled to the places where the majority of poor lived, through *integrated rural development* programs – or rural-oriented service and production projects; (2) aid was focused on overcoming deficiencies in the living standards of the poor, through *basic-human-needs* programs; and (3) aid was focused on groups with common socioeconomic characteristics that created or maintained them in poverty, through projects designed for *special publics.*

Integrated Rural Development

One means of reaching the poor majority was to channel assistance to those places where the poor lived. The World Bank's (1975: 5) rural-sector-policy paper asserted that "a special effort is needed to provide appropriate social and economic infrastructure for the rural poor and it is important to integrate these components into rural development projects." Without a concerted effort, the Bank staff insisted rural poverty would remain pervasive. By providing "minimum packages" of inputs, organizing area development programs, and combining sectoral investments in rural areas, the World Bank sought to: increase agricultural productivity; expand off-farm employment; and increase entrepreneurial opportunities for the rural poor.

Much like the World Bank's lending strategy, USAID's program of assistance sought to increase the productivity and income of the rural poor, and to extend access to services and facilities to rural families. USAID officials, responding to the new mandate from Congress, recognized in the early 1970s that traditional forms of aid for roads, irrigation, public works, and rural electrification, while necessary, were far from sufficient to increase the productivity and income of the poor. During the 1960s such investments had primarily helped large commercial farm owners but had done little to extend the benefits to landless laborers, tenant farmers, small-farm holders, and the rural unemployed – who became the primary beneficiaries of USAID programs in the 1970s. USAID's (1973) Working Group on the Rural Poor argued that the more traditional projects had to be redesigned to reflect the needs of these new beneficiaries and should be combined with programs for human resource development, education, health, family planning, small-scale industrialization, and labor-intensive agro-processing. The access of small-farm holders to appropriate technology, new production inputs and markets for their products had to be increased. The Task Force on the Rural Poor insisted that USAID's strategy include programs for creating and strengthening local institutions – such as cooperatives, farmers organizations, municipal and district governments and regional planning units – to facilitate and coordinate service delivery.

Moreover, USAID strategists saw the relationships between urban and rural economies in a new perspective. They called for projects that would create or strengthen economic linkages between rural villages, market centers, smaller cities and regional urban centers to integrate the settlement and spatial systems of developing regions. They pointed to the need for projects that would contribute to the "creation of market areas and market towns complete with services and amenities designed to make rural life productive and satisfying" (USAID 1973: 17).

USAID, the Canadian International Development Agency, UNDP, and the World Bank funded integrated rural and regional development (IRD) projects worldwide. The German aid agency, GTZ, focused its IRD projects on building both national and local organizational capacity to plan and implement capital and service delivery projects through technical assistance and training. NORAD, the Norwegian assistance organization, carried out IRD projects in Southeast Asia and Africa that focused heavily on training and the development of local institutional capacity. British funding in Anglophone Africa emphasized the development of district planning and management capabilities to provide rural infrastructure and services needed for rural economic development. SIDA, the Swedish aid agency, focused its efforts on helping African governments decentralize their operations and build the capacity of local organizations to support agricultural production, social services, and local revenue raising (Conyers *et al.* 1988).

However, the experience with integrated rural development projects and with developing institutions to provide services to farmers produced mixed results. In a review of 200 rural development projects financed by the UNDP and executed by United Nations specialized agencies during the 1970s, UNDP (1979) evaluators found that many agricultural and rural development activities were not successful because those who designed the projects limited external assistance to strengthening institutions at the central government level and paid little attention to local organizations.

Many of the technical cooperation projects undertaken by United Nations agencies during the 1970s to develop institutional capacity to carry out rural development programs faltered because they were not tailored to local conditions and needs. A review of sixty UNDP-funded projects showed that they fell short of their objectives because assistance resources were "used to fill shortfalls in technical and administrative capacity rather than to inculcate new skills or to test new ideas and approaches to rural development" (UNDP 1979: 8). The most frequently reported problems were the failure of project designers to identify the socioeconomic and physical environmental factors that would adversely affect beneficiaries and a tendency for project designers, government officials, and technical assistance experts to view local communities as passive rather than active participants. National officials and international experts tended to transfer technology directly from more developed countries, often with only limited consideration of the conditions or perceptions of the rural population.

The faulty design of integrated rural development projects was attributed primarily to the way in which the UNDP perceived the projects. As the UNDP evaluation pointed out:

> Design suffers because UNDP projects are regarded by project personnel almost as self-sufficient entities in which the importance of international inputs is paramount . . . Project inputs are not adequately linked conceptually with the institution or government program which they are meant to support. Little attention is given to the aims and design of the overall national program, so that the relationship between the project support and the program is difficult to determine. The project becomes an end in itself.
>
> (UNDP 1979: 24)

Studies of IRD projects that were sponsored by USAID and the World Bank (see Honadle *et al.* 1980, Ayres 1984) found a number of common and frequently recurring problems, including: difficulties in integrating and coordinating the activities of the many participating government agencies required to provide agricultural, social, and productive services; difficulties in managing and supervising teams of multidisciplinary technical and administrative staff needed to carry out the projects; and inadequate information to make effective managerial decisions. Most IRD projects lacked adequate incentives to obtain the support and commitment of project staff or cooperating organizations. Project managers in remote rural areas frequently had problems obtaining supplies, equipment, and personnel in a timely manner to carry out the project on schedule, resulting in delays and cost overruns and the diversion of resources intended for integrated rural development projects to other activities. Few of the benefits of IRD projects were sustained when foreign assistance ended.

Basic Human Needs Strategies

A second approach to channeling aid to the poor emerged in the mid-1970s. United Nations specialized agencies, such as the International Labor Organization, began calling for a new emphasis on employment-generating projects that would help the world's poor obtain higher standards of living and basic human needs. The World Employment Conference, organized by the ILO in 1976, argued in its Declaration of Principles that previous development strategies in most developing countries had not led to a reduction of poverty and unemployment, and that major shifts would be needed in both national and international development strategies to provide full employment and adequate income for the poor as quickly as possible (ILO 1976).

The ILO defined basic needs to include two components: minimum family requirements for consumption such as adequate food, shelter, clothing, household equipment, and furnishings; and essential community services such as potable water, sanitation, health services, educational facilities, and public transport. The ILO insisted that policies for satisfying basic needs be accomplished as much as possible through self-reliant development and internal resource mobilization.

The objectives of basic-needs strategies were defined in two ways. The World Bank and other assistance organizations saw the provision of basic goods and services as a precondition for increasing the productivity and income of the poor, enabling them to contribute more effectively to national development. Others, such as Burki (1980: 18), argued that the aim was to "ensure the access of the poor to a bundle of essential goods and services" as a basic human right. The distinctive features of a targeting strategy for meeting basic needs in the latter sense were outlined by Streeten and Burki (1978) as follows:

- it is concerned with meeting the needs of the poor as a legitimate goal aside from its contribution to productivity
- it stresses the importance of alleviating absolute poverty
- it emphasizes the need for supply management so that the increased income of the poor is not offset by rising prices for basic goods and services that they must purchase
- it emphasizes the restructuring of production so that the poor have greater access to basic goods and services despite their disadvantages in the market
- it is defined by characteristics of the goods and services needed by the poor rather than in terms of the goods themselves (nutritional needs rather than specific types of foods)
- it seeks to divorce production decisions directly from market-based consumers" choices in situations where income distribution is extremely uneven
- it encompasses a variety of non-material needs, such as sense of purpose in life and work, that are often satisfied as the result of obtaining material needs.

Recognizing the need to improve the health of the rural population in order to increase labor participation and productivity, both UNDP and USAID sought to strengthen primary health care for the poor. United Nations programs funded through UNDP, the World Health Organization (WHO), UNICEF and the Fund for Population Activities, focused on strengthening ministries of health. Often, however, these programs had elements for strengthening the capacity of other public institutions, private voluntary organizations, religious or church groups, and private-sector organizations to deliver primary-health-care services (UNDP 1983). Much of the assistance was for education, training, and skills upgrading of health-care workers.

Evaluations of United Nations projects in primary-health-care improvement found that their outcomes were affected by a variety of factors. Among the factors making improvements difficult were "environmental" problems – that is, social and economic conditions that were not conducive to institutional development in the primary-health-care sector, including inadequate numbers of primary-health-care personnel, inappropriate training, lack of resource planning for health, and lack of continuing education for health-care staff. Management problems also limited the effectiveness of the projects – they included inadequate teamwork and supervision of health workers, inadequate salaries and incentives, lack of job descriptions, absence of relationships with traditional healers, low motivation of personnel. Weak community participation and low levels of service use undermined efforts to improve service delivery. Many of the problems were caused by: the low priority that national governments gave to primary health care; the meager allocations to health care from the national budget; the lack of political commitment in many countries to improving health services (UNDP 1983).

Many socialist governments also focused their development programs on a basic-needs strategy. China's development program during Mao's regime, for example, aimed at: providing for basic needs; creating a "floor" of income under which families would not fall; and increasing access to productive inputs through communal organization of agriculture and small industry. In this aspect of development, the success of China and some socialist countries (such as Burma and Sri Lanka) provided evidence that basic-needs approaches could be effective in alleviating or eliminating absolute poverty. They provided less impressive evidence that basic-needs strategies alone could increase productivity or stimulate enough economic growth to sustain them (Paine 1976, Lee 1977, Prybyla 1979).

Assistance to Special Publics

A third approach to channeling aid to the poor majority emerged in the mid-1970s from the realization that those living in poverty were heterogeneous groups with special characteristics, such that their problems would not be alleviated entirely by general poverty policies. Unless the productivity and income of these groups were raised, overall economic growth would be difficult to attain. World Bank (1980) officials pointed out that the high levels of skill and education in industrial countries were a cause as well as a result of economic growth. Developing nations, they contended, must be able to incorporate the poor with unique characteristics into productive activities – through special educational, health, and social-services projects. "People who are unskilled or sick make little contribution to a country's economic growth," World Bank (1980: 36) analysts noted. "Development strategies that bypass large numbers of people may not be the most effective way for developing countries to raise their long run growth rates." They underlined the need for programs and projects tailored specifically to the needs of particular groups of the poor to ensure their participation. In their extensive studies in South Asia, Griffin and Khan (1978) found that "poverty is associated with particular classes or groups in the community; e.g., landless agricultural laborers, village artisans, plantation workers, etc." Yet, most of the development policies and programs that had been proposed previously were "couched in terms of atomistic households in a classless society. This neo-classical assumption," they contended, "is closely associated with an assump- tion of universal harmony of interests."

But experience with development policies aimed at alleviating poverty continued to demonstrate that their success was often jeopardized by conflicts among groups with very different interests and that many groups living at subsistence levels had characteristics that continued to exclude them from benefits of integrated rural development and basic-needs programs. Esman and Montgomery called these groups "special publics" and described the circumstances that made their participation unlikely:

> These special publics often live in remote or hard-to-reach areas or suffer their greatest privations during the wet season, when they cannot be approached by ordinary overland routes; a few even make it a point not to be seen at public facilities set up for supplying family planning and health services, nutritional supplements, or even primary nonformal education.
>
> (Esman and Montgomery 1980: 190)

Government bureaucracies that administered programs for the poor were rarely organized and equipped to deal effectively with these special publics and, as Esman and Montgomery (1980: 190) noted: "for their part, the end-of-the-line field workers in a human development service are rarely motivated to break the cognitive, social and physical barriers that separate them from the special publics." The supply lines and flows of information to those working with special publics were continually interrupted and, because these groups lacked political power, their needs and demands were rarely considered in the formulation and implementation of development policies in the national capital. The only effective way of reaching these groups was to design projects that in some way were tailored to their specific needs and accounted for their special characteristics, a task with which international agencies and governments in developing countries have yet to come to grips.

PROJECT PLANNING AND DESIGN PROCEDURES

Just as the underlying concepts of development theory and strategy were being reassessed during the 1970s, so were the procedures by which development projects were planned and implemented. International assistance agencies and most governments of developing countries during the late 1960s and early 1970s attempted to plan and control development projects through complex design, selection, and appraisal procedures. The criteria to which project proposals were subjected were seen in the "project cycles" that guided international agencies" activities. The World Bank's (1990) cycle, for instance, consisted of six stages through which all projects had to proceed:

1 *Identification*: Borrowers, Bank staff, or other international organizations identify suitable projects that support national and sectoral development strategies and are feasible according to Bank standards. These projects are then incorporated into the Bank's lending program or "pipeline" of projects for future consideration.
2 *Preparation*: The borrowing country and Bank staff develop the initial project idea into a detailed proposal, considering its technical, institutional, economic, and financial aspects. The Bank provides guidance and helps the borrower obtain financial and technical assistance with design, feasibility studies, and other aspects of preparation.
3 *Appraisal*: Bank staff review comprehensively and systematically all aspects of the project . . . and cover four major aspects: technical, institutional, economic and financial. An appraisal report is prepared on the return of Bank staff to headquarters and is reviewed extensively. This report serves as the basis for negotiations with the borrower.
4 *Negotiations*: The Bank holds discussions with the borrower on the measures needed to ensure success for the project. The agreements reached are embodied in loan documents. The project is then presented to the executive directors of the Bank for approval. After approval the loan agreement is signed. The project can now go into its implementation stage.
5 *Implementation and supervision*: The borrower is responsible for implementation of the project that has been agreed with the Bank. The Bank is responsible for supervising that implementation, through progress reports from the borrower and periodic

field visits. An annual review of Bank supervision experience on all projects underway serves to continually improve policies and procedures. Procurement of goods and works for the project must follow official Bank guidelines for efficiency and economy.

6 *Evaluation*: An independent department of the Bank reviews the completion report prepared by the Bank's projects staff, and prepares its own audit of the project, often by reviewing materials at headquarters or examining project documents and results in the field. This ex-post evaluation provides lessons of experience that are supposed to be built into subsequent identification, preparation and appraisal work.

Although these cycles were relatively simple iterative planning models that had an internal logic and could be used effectively as guides to action, the problems of applying them arose from the complex requirements that were often attached by the international assistance organizations (see Fig. 3.1).

Project preparation was preceded by the identification of an idea for an investment, either through formal or informal processes. In most developing countries, proposals emerged from a variety of sources: through unguided entrepreneurial investment, from programs of government ministries or agencies, or through proposals submitted by private investors for government sponsorship or participation (Rondinelli 1976b). But regardless of their source, once an idea was submitted international assistance agencies insisted that proposals be prepared in sufficient detail to test their feasibility using a complex set of financial, economic, technical, and administrative criteria.

For example, the United Nations Development Program (UNDP) required preinvestment or prefeasibility studies: to determine the most appropriate means of defining an initial proposal; to gather supporting data to justify the proposal before various government agencies and review committees within UNDP; and to collect information required later for preparing a prospectus and in conducting more detailed feasibility and appraisal analyses. Generally, the UNDP suggested that preinvestment studies should also identify potential bottlenecks, obstacles, and preconditions for successful execution of the project and should recommend the administrative and policy reforms required to implement it (Taylor 1970, Ewing 1974).

Depending on the form of the request and the type of project under consideration, UNDP preinvestment studies were either

Project identification

- Perform macroanalysis: country programming, sectoral planning, forecasting and projection, policy analysis
- Perform microanalysis: gap or needs analysis, input–output studies, capital and deficiencies analysis, social analyses
- Set immediate and long-range development goals
- Set specific development targets for sectors
- Identify potential programs and projects

Initial reconnaisance and project definition

- Establish need or demand for proposed project
- Define linkages with development plans
- Determine potential 'external effects' of project proposal
- Identify beneficiary groups or target areas
- Estimate 'order of magnitude' costs
- Estimate other resource commitments
- Seek initial political and administrative support
- Circulate 'idea stage' proposal to technical and operating ministries for review

Prefeasibility analysis and formulation

- Refine project objectives and targets
- Define potential components
- Evaluate potential design configurations
- Determine appropriate project size and potential location
- Analyze preconditions for successful implementation
- Estimate potential costs and benefits
- Prepare suggested financing plan
- Refine justification analysis
- Secure preliminary government review and approval
- Obtain preliminary review by potential funding sources

Project design and prospectus preparation

- Determine major local conditions and needs, constraints and parameters affecting design
- Delineate components, activities and tasks
- Prepare blueprints and specifications for equipment and facilities
- Prepare initial operating plan and work schedule
- Prepare life-of-the-project budget
- Prepare detailed job descriptions
- Estimate technological requirements and plan for technology adaptation
- Establish preliminary resource acquisition plans
- Prepare preliminary production and delivery schedule
- Choose appropriate organization for implementation
- Prepare formal prospectus

Formal feasibility analysis

- Perform detailed market studies
- Perform commercial, financial and economic studies
- Perform site and location studies
- Perform technical and technological studies
- Assess organizational and managerial arrangements
- Analyze social and environmental impacts
- Calculate least-cost or cost–benefit estimates
- Review operational plans
- Assess administrative capability of sponsoring agency
- Recommend changes in project design

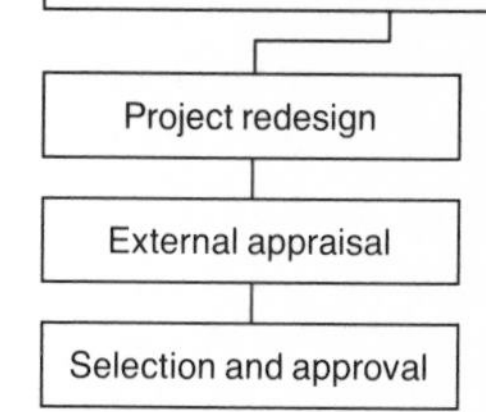

Figure 3.1 Major requirements for development project presentation

general investigations of sectoral or national development needs that yielded potential ideas for project proposals or studies of the best way of formulating an initial idea for a project. If an idea had already been formulated into a preliminary proposal, the preinvestment study would involve more detailed analyses. In the prefeasibility assessment of potential agricultural projects, for example, the UNDP required information on the volume and value of expected incremental production based on existing prices; on markets for commodities; on storage, transport, processing and distribution facilities; on price and cost incentives to producers, processors and distributors; and on the likely costs and benefits of the project. Similarly, for potential housing projects the UNDP wanted information about the housing market – housing requirements within the country, effective demand and supply, distribution of income, the propensities of potential buyers to save, available private financing, and government financing and subsidy policies (UNDP 1973).

USAID required from its field missions a project identification document (PID) for each proposal, which served as a preinvestment analysis by providing: information on how the proposal related to USAlD's assistance policies and priorities in the country; national and sectoral development objectives; a description of the beneficiaries; and the key social and economic factors that might affect the success of the project.

International assistance agencies and governments in developing countries also attempted to control the design and implementation of projects by requiring sponsors and field staff to submit detailed prospectuses in the project formulation and preparation stages. The UNDP (1974), for instance, used a standard "project document format" requiring a concise statement of long-range sectoral or national development criteria and immediate project objectives defined in terms of verifiable results. Project activities – the tasks that had to be performed to achieve immediate objectives – were to be described in detail. Project inputs – "the resources which must be made available at specific times, in specified quantities, and in accordance with agreed specifications" – also had to be identified. The documents had to describe: the institution that would execute the project, and its purposes, programs, financial and staff resources, and organizational structure, as well as its relationship to other institutions; the legal framework in which the project would operate; coordination arrangements; and provisions for government follow-up activities. Mid-project review and terminal evaluation procedures, as well as plans for subcontracting, training, procurement of equipment and supplies, provision of government contributions, and operations and maintenance, all had to be specified. Each project document had to contain a detailed budget that displayed costs for all inputs as well as the expected sources of revenue. Depending on the size of allocations requested, the document was reviewed by both the UNDP resident representative in the developing country and headquarters staff in New York.

Similarly, USAID required for each project a set of increasingly detailed documents: initially a description of potential ideas for inclusion in Congressional budget submissions and later a project identification document that set out the proposal in brief and identified the resources required to design it in more detail. Finally, a full-scale proposal, in the form of a project paper, had to be submitted. For each project submitted by field missions USAID required a project design summary or "logical framework" consisting of four sets of data:

(1) the goals or objectives of the project, and the program or sector goals to which the project would contribute, including the project's specific purpose, the intended outputs delineated, and the required inputs or resources identified;
(2) objectively verifiable indicators, including measures of goal achievement, conditions that would indicate at project's end that the purpose had been achieved, the magnitude of outputs, and type and quantity of implementation targets;
(3) means by which the indicators could be verified; and
(4) important assumptions concerning the ability to achieve goals and targets, project purposes, outputs, and the means of providing inputs.

The logical framework required project designers to deal systematically with management issues such as priority setting, programming, budgeting, and control.

The project paper also had to assess the project's technical, environmental, financial, and social soundness, and review cost and benefit implications and implementation requirements. The implementation analysis included:

(1) a description of the recipient government's administrative arrangements for implementing the project and an assessment

of the implementing agency's potential for providing leadership, resources, commitment and "grass roots" management;
(2) an implementation plan that outlined responsibilities for actions within the implementing agency and described its relationship with other government organizations;
(3) a network chart that showed the scheduling of activities over time and identified milestones for measuring success or progress;
(4) a description and discussion of potential problems or weaknesses about which reforms or policy changes would be negotiated with the government;
(5) a description of proposed monitoring and evaluation techniques;
(6) an analysis of logistical support required for implementation; and
(7) a description of evaluation arrangements including plans for host government collaboration in evaluation, collecting baseline data, supervising progress, and training evaluators.

Moreover, each international assistance agency reviewed project proposals using its own feasibility and appraisal procedures, and often required modifications in design as a result of the findings. International assistance organizations prescribed a complex set of requirements for appraising development projects.

The World Bank strongly influenced the design of projects through feasibility and appraisal analyses, through formal and informal negotiations with the potential borrower before and after appraisal, and through bargaining on the final loan agreement. Bank officials, attempting to exercise "leverage" on the financial and administrative procedures of borrowers, insisted that preliminary analysis and direct technical assistance were important instruments for convincing governments to modify or change design proposals. The Bank's policy handbook declared that:

> By assisting government in the formulation of development programs and by working with borrowers to prepare high priority projects for external financing, it ensures that by the time a formal request for financing is submitted, many critical issues will already have been raised and agreements reached.
>
> (World Bank 1974: 46)

The Bank was able to modify projects to better fit its own priorities, reduce costs, increase efficiency, or improve financial or organizational features prior to final negotiations. The extent of the Bank's influence on design was only hinted at by the statement that "sometimes a substitute project or one conceived on a somewhat different scale has been found to be more productive than the one originally proposed."

The World Bank used a comprehensive appraisal procedure to evaluate project proposals (Baum 1978). Appraisal reports were usually prepared by a team of Bank economists, sometimes with assistance from outside technical experts, and included assessments of six major elements:

(1) *economic aspects*, including the strength of the sector in which the project was to be undertaken, the demand for goods and services that would be produced by the project, descriptions of costs and benefits, calculation of a rate-of-return, analysis of the effects of the project on balance of payments, cost analyses of potentially feasible alternatives, and least-cost analyses of non-revenue producing projects;
(2) *technical aspects*, especially the engineering feasibility, appropriateness of scale and location, adequacy of manpower to prepare and supervise the project, evaluation of proposed methods and processes, layout and facility design criteria, construction scheduling, and environmental consequences;
(3) *managerial aspects*, including the quality of management, availability of managerial talent and experience, and the need for management agents;
(4) *organizational aspects*, with an assessment of the appropriateness of organizational structure for constructing and operating the project, issues of centralization versus decentralization of authority and responsibility, internal management capability in budgeting, programming and reporting, control and maintenance capacities, and adequacy of training programs;
(5) *commercial aspects*, particularly the adequacy of procurement systems, availability of inputs for production, and marketing and distribution channels; and
(6) *financial aspects*, including an overall evaluation of financial soundness as indicated by value of fixed assets and inventories, terms of existing debt and sources of capital, review of start-up and operating costs, and the need for and sources of new capital.

The Bank and most other assistance agencies could also insist on drastic changes through formally and informally negotiated agreements with the recipient government. These "conditions of effectiveness" for World Bank loans or "conditions precedent" in USAID project agreements required the government to make changes in the project's components, scope, and organization prior to, and as a condition of, disbursement of loan or grant funds.

Negotiated agreements could be quite detailed. In a loan for a rural development project in one African country, for instance, the World Bank Group's International Development Association (IDA) insisted on conditions that specified the number and types of wells and pumps to be used in irrigation systems, the structure of the organization that would finance loans by the nation's rural development bank, and the creation of a separate management unit to execute the project. IDA also insisted on centrally consolidated accounts for the rural development fund. It set guidelines for the types of administrative and technical personnel to be hired, and prescribed mechanisms for providing extension and credit services to farmers as well as procedures for performing cost–benefit analyses in sub-projects. In addition, IDA established detailed duties and responsibilities for the project manager and outlined the contents of contracts for farmers participating in the sub-projects. The government was required by conditions of the loan agreement to form an interdepartmental technical committee to coordinate project activities. As the consequences of these complex planning procedures and management controls became more obvious during the 1970s, many critics began questioning their value. As their own supervision, monitoring, and evaluation reports began to indicate, international assistance agencies created a system for project administration during the 1960s and 1970s that was beyond their capacity and that of developing nations to implement effectively. The tendency to abstract, rationalize, standardize, control, and complicate not only created the conditions for failure but inflicted hardships and frustrations on the intended beneficiaries (Rondinelli 1987).

Observing the trend toward more sophisticated analysis and detailed planning for rural development projects in Africa, Chambers and Belshaw concluded that:

> together these lead away from reality, from what is feasible, and the cumulative increments of change which can gradually transform performance, and encourage the design and propagation of ideal models which are not only unattainable but also liable to impair rather than improve performance.
>
> (Chambers and Belshaw 1973: 6.3)

Assessing the impact of increasingly complex procedures on project administration, they argued that "the perfectionist planner and the intellectual academic are both susceptible to recommending yet more planning, more detailed and specific statement of objectives, the generation and analysis of more data, the identification, elaboration, and choice between more alternatives." The trend toward more complex analysis, they concluded, had two unfortunate results: "generating an insatiable appetite for planners, who are far from costless; and reducing the chances of anything happening on the ground."

To many officials in developing countries, adoption of foreign planning procedures and management systems served no rational purpose other than to obtain external funds. Complex appraisal and management procedures often could only be applied by foreign experts, academics, and consultants, with no guarantee that the analyses would beneficially alter the substance of a proposal, that the project would be more palatable to local bureaucrats or politicians, or that the projects would be better implemented within a traditional organizational and political structure. Many of the prescriptions were simply irrelevant to the conditions under which government officials had to operate; or worse, they conflicted with internal political processes required to obtain support and approval for project proposals.

Obstacles to systematic project planning and implementation, as will be seen in Chapter 4, pervaded bureaucracies in developing nations. Unless high-level political support protected a proposal through its early stages, then rivalries, conflicts, and bureaucratic competition delayed or killed even "high-priority" projects justified by voluminous feasibility studies. Politicians in strategic positions within central ministries or provincial governments often pushed through pet projects that were badly formulated and lacking feasibility studies, displacing those delayed by intensive analysis and appraisal. Told by international experts to define goals and objectives clearly, government officials knew that it was frequently to their own advantage to keep objectives and purposes "fuzzy" and amorphous, to avoid unequivocally stating intended accomplishments, and to resist losing flexibility by submitting detailed designs and operating plans. In an atmosphere of uncertainty and resistance to change, success depended on obtaining and preserving a strong coalition of political support. Projects survived by attracting advocates with different goals, purposes, and expectations, each seeking different outputs and "pay-offs." Objectives were better left vague, allowing each reviewer to form his own conclusions, than made explicit at the risk of generating conflict and opposition or of losing flexibility during implementation.

While economists, planners, and management analysts in international assistance agencies devised objective and rational allocation and appraisal techniques, sophisticated selection procedures, and systematic scheduling and control devices to optimize resource allocation and utilization within the national economy, government officials were concerned with the political compromises and trade-offs necessary to push any project through a politically influenced selection process. Moreover, problems arose even if a project's sponsors wanted to define objectives precisely and relate them clearly to national development policies. Goals, targets, and priorities were kept vague in project proposals for the same reason that objectives of national development plans had been ill-defined: to obtain consensus and minimize political conflict. Goals changed quickly in developing nations, policies were displaced by succeeding regimes. Those developed for external consumption often had low internal priority and those espoused by ministry heads were often publicly acknowledged by national political leaders but given little budgetary support.

Even if all parties could agree on precise goals, priorities, and desired results, few developing countries had reliable statistics, adequate data, and skilled analysts to apply the procedures prescribed by international funding institutions. For example, a USAID education-sector analysis team sent to Brazil found it nearly impossible to apply the prescribed methods in one of the more economically advanced developing countries. Neither the education ministry nor the state governments had the data or trained manpower to apply sophisticated analytical techniques. The mission reported that:

> The pattern of financing of education in Brazil makes it difficult to effectively analyze the cost efficiency of the education sector or the precise future financial needs of the system. It is very difficult to satisfactorily plan the expansion of education programs or to know to what extent the effort to increase productivity of existing resources might have a significant impact on decreasing per pupil costs. Likewise it is impossible to evaluate the cost-benefits of investments in innovations of educational technology if costs of present educational methods are not known.
>
> (USAID 1970: 6)

Perhaps the ultimate irony was that many developing countries had been judged to have backward and inefficient administrative systems because they could not apply analytical and management techniques, the efficacy and practicality of which remained unproven even in western industrial nations.

By the early 1970s, a number of the critiques of rationalistic planning and analysis procedures reviewed in Chapter 1 were emerging in the United States to question the usefulness of systems analysis, quantitative models, and planning/programming/budgeting schemes in public decision-making (Wildavsky 1969, Schick 1973, Hoos 1972). Studies of weapons-systems projects in the United States, from which many of the planning and management procedures had been borrowed, indicated that PERT and other scheduling and control techniques gave a "myth of managerial effec- tiveness" to project administration without contributing to sub- stantive results (Sapolski 1972). An extensive analysis of factors determining project success commissioned by the US National Aeronautical and Space Administration (Murphy *et al.* 1974: 9) found that "creation of elaborate and detailed reporting and control systems . . . detract from success by causing excessive delays, red-tape, super- ficial reports and inadequate information flows."

Thus, by the late 1970s both strategies of economic development and procedures and techniques of project planning and implementation again came under increasing scrutiny.

STRUCTURAL ADJUSTMENT: A RETURN TO MACROECONOMIC GROWTH POLICY

The evaluations of international assistance and development policies that were carried out during the late 1960s and early 1970s indicated that neither growth maximization and trickle down, nor semi-targeted sectoral development policies, were sufficient to overcome the growing disparities between rich and poor countries or to stimulate economic growth with social equity in developing nations. Even where conventional strategies had been successful in promoting growth of GNP, they often did little to meet the needs of the majority of the poor living at or near subsistence levels who had previously been excluded from participation in economic activities. The numbers of people living in dire poverty were increasing rather than diminishing. In countries that had achieved respectable levels of economic growth with relatively equitable distribution, substantial efforts had been made prior to their period of growth to satisfy basic human needs, redistribute productive assets and intervene in market processes to assure widespread distribution and access. By 1973, international aid agencies were not only rethinking their policies and strategies, but also reformulating basic theories of development and fundamental goals of international economic assistance. The need for more precise development policies and for more effectively targeted aid strategies that would reach specific groups of the poor became more widely accepted by the late 1970s.

Ironically, however, the strategies of international assistance organizations – led by the World Bank – shifted in the early 1980s from programs to increase the income, productivity and living standards of the poor to those for macroeconomic restructuring. Structural adjustment policies were aimed at helping developing countries cope with their severe debt problems by stimulating rapid economic growth. They sought to overcome problems associated with severe international recession and inflation, declining demand for exports, rising prices for imports, and decreases in international financial assistance. The 83 percent increase in the price of oil in 1979 and 1980 alone placed extreme burdens on the budgets of petroleum-importing countries.

Initially, the World Bank justified the shift in emphasis by pointing out that these new costs, along with substantial decreases in the flows of international assistance, threatened the survival of human development projects and curtailed the ability of governments in most countries to address the problems of widespread poverty. The economic trends of the early 1980s led international economists to conclude that the "outlook for reducing poverty has worsened along with the prospects of the poor countries" (World Bank 1981: 18). About 750 million people (excluding China where accurate estimates were not available) were living in absolute poverty in 1980. Even under the most optimistic economic conditions – with high rates of population growth – that number could realistically be expected to decline only 18 percent by the end of the twentieth century, still leaving more than 615 million people living at the subsistence level. With sluggish economic growth, the number of desperately poor could increase to 850 million.

Thus, by the early 1980s, the attention of governments in developing nations was refocused on what the World Bank considered to be three of the most important influences on their ability to survive the ensuing international economic crisis: trade, energy and external finance. World Bank analysts argued that the capacity of developing countries to adjust to changing economic conditions would depend on:

> their ability to increase the flow of external capital and to raise the rate of domestic saving in order to finance investment aimed at restructuring their economies. Also of critical importance will be their success in increasing export growth and reducing dependence on imported oil, capital goods and raw materials.
>
> (World Bank 1981: 14)

The shift in concern among international organizations from alleviating poverty to promoting macroeconomic growth also refocused their perspective on development – from a bottom-up to a top-down approach. The World Bank (1983) moved from the equity concerns that it had been pursuing for a decade to reforming monetary, fiscal, exchange-rate, trade, and wage policies. World Bank structural adjustment loans (SALs) provided funding to support a comprehensive program of economic reform agreed to by the government and the Bank.

Although some countries were able to adjust their economies in order to service their debts effectively during the early 1980s, many of the non-oil-producing and poorest countries faced serious difficulties and were cut off from further borrowing by private capital markets. According to Krueger (1989: 8), the inability of these countries to cope with debt repayment convinced multilateral lending institutions that economic policies were the primary determinants of their overall economic growth rate and that, therefore, "these policies cannot be ignored in assessing the probable future rates of return on new lending, be it project or program-based."

The solution to the debt-servicing problem was defined by the IMF and World Bank as drastic economic reforms that would accelerate the growth of GNP. The reforms would promote open trade, export production, deregulation of industry, reduction of public expenditure and the public wage bill, and expansion of private-sector output. The debt crisis provided an opportunity for neo-classical economists to reassert their influence in the international lending institutions and to reorient development assistance strategies. Their approach, according to Krueger (1989), was based on several assumptions. First, they believed that international lending institutions did not have sufficient financial resources to displace private capital markets. Rather, their comparative advantage was in using their resources as "leverage" to change economic policies and thereby make developing countries more creditworthy. In the long run this would have a greater impact than merely lending more money to already debt-ridden countries. Second, they saw little reason to support projects in countries where economic policies were so inimical to growth that rates of return would inevitably be low. Third, most middle-income developing countries did not need project assistance, and the multilateral agencies could better provide support for economic reform. In those countries with substantial debt problems, a higher return on international resources would come from investment in structural adjustment than from individual development projects. Finally, after experiencing a decade of frustrations with attempting to promote social change, they saw a more important role for international lending institutions in creating a healthy international economy in which developing countries would grow through trade and investment.

The structural adjustment and neoclassical economic policies of the 1980s and early 1990s were no more the panaceas for development problems and no less criticized for their limitations, however, than the policies that preceded them. Even the harshest critics of macroeconomic adjustment policies admitted that the economies of many developing countries required fundamental readjustment and reform. Their arguments with this strategy were based on the problems that arose from the way in which structural and sectoral adjustment loans were planned and implemented. Even World Bank President Robert McNamara, who thought that failure to deal with the international economic crises would irreparably harm developing countries, admitted that "human development is threatened during the adjustment period and the potential consequences in unnecessary human suffering are grave" (World Bank, 1981: iii).

Critics argued that although experience suggested that redistribution without growth could not long be sustained (Adelman and Morris 1967), the renewed emphasis on macroeconomic growth took a narrow view of development. Economic growth without attention to distributive issues and human resource development historically yielded little in the way of progress for the poor (Adelman 1986).

By the mid-1980s studies did indeed begin to show that IMF conditionality and World Bank structural adjustment lending had mixed impacts on the economies of developing countries and negative effects for the poorest populations (Helleiner 1987, Demery and

Addison 1987). Some of the most serious problems arose because international assistance organizations paid little attention to the social consequences of the prescribed reforms. In many developing countries, adjustment policies resulted in higher levels of unemployment in urban areas and lower incomes for the poorest segments of the urban labor force. In countries like Costa Rica, where structural adjustment policies were quite successful in restoring an economic environment conducive to export promotion and private-sector investment, unemployment eventually dropped, income levels rose and poverty was reduced (Fields 1988). But in other Latin American countries, where the policies were either inappropriate or ineffectively implemented, they adversely affected national labor markets (Corbo and de Melo 1987).

Adjustment policies adopted in most Latin American countries were biased against lower-income urban workers who could least afford the cost of adjustment. In the Southern Cone countries of Latin America, for example, economic stabilization and liberalization policies worked at cross purposes, creating adverse effects for national labor markets (Corbo and de Melo 1987). "This recessive policy package generated a contraction in the demand for labor by modern firms which, given the rapidly increasing labor supply, produced a significant expansion of open unemployment," Tokman (1988: 119) noted. Not only were there fewer new jobs, but those that were created were of lower quality. Most Latin American countries saw rapid increases in informal employment, in service employment, and in public employment, along with decreases in private-sector productivity and income. Manufacturing employment in Latin American countries declined on average by 2.2 percent a year between 1980 and 1985. During the same period wages declined in construction (–3.3 percent), industry (–2.3 percent) and in the public sector (–3.2 percent).

In Asia, the poorest participants in the urban labor force suffered from both the disease and the cure: first from the adversities of recession and lower rates of economic growth, and later from the adverse consequences of structural adjustment and fiscal austerity policies. The International Labor Organization (ILO) pointed out that structural adjustment policies in many Asian countries left the poor worse off because they reduced incomes, increased unemployment, and raised the prices of food and other essentials. The ILO analysts also noted that:

> In the worst affected country, the Philippines, large numbers of people, including modern-sector wage employees have been exposed to loss of real income . . . In most countries, it is those who are outside the modern or organized sector labor market who have suffered the worst setbacks.
>
> (ILO–ARTEP 1987: 89)

Those who designed structural adjustment policies often ignored or discounted the political and institutional requirements for policy implementation. The IMF often forced its austerity programs on political leaders in developing countries, many of whom retrenched in the face of violent opposition or simply ignored the policy changes needed to carry out the reforms (Bacha 1987). The record of implementing World Bank sectoral adjustment loans (SECALs) was also mixed. Reviewing the results of fifty-five such programs carried out between 1983 and 1987, Paul (1988: 7) concluded that "in general, [only] the relatively simpler institutional reforms were implemented; the more complex and politically sensitive reforms faced severe problems."

Many of the IMF's austerity programs and the World Bank's structural adjustment loans (SALs) tried to cut back drastically the role of government in the economy. But in the rush to reduce the size of government in developing countries, the International Monetary Fund and World Bank overlooked the critical roles government played in western economic growth and in the success of the East Asian economies. Neoclassical economic theories ignored the fact that not even the most successful industrializing Asian countries relied entirely on free operation of the market. Singapore, South Korea and Taiwan, for example, depended heavily for their economic growth on the export production that structural adjustment policies envisioned, but the central government played a catalytic role in providing the public services and infrastructure needed for economic production, in breaking the "bottlenecks" to economic expansion, in investing directly in productive activities during the early stages of economic growth, and in creating favorable conditions for private enterprise development in later stages (Rondinelli and Montgomery 1990).

South Korea's experience in pursuing rapid economic growth underlined the strong role that the state played in mobilizing and guiding both public- and private-sector organizations during the 1970s. It succeeded initially by using state investments and coercion, and later by relying primarily on incentives and guidelines enforced through state regulations. Moskowitz (1982: 71) points out that "here is where goal focus, political will, executive dominance, bureaucratic commitment, and the 'hard state' combine ingeniously with Korean entrepreneurial talent to achieve the desired end, economic growth."

Structural adjustment policies were usually complex blueprints for economic reform, and international organizations imposing them often overlooked the fact that long-term success in development depended less on a particular set of policy reforms than on the expansion of institutional and administrative capacity to adjust to changing economic conditions. Evaluations of experience with macroeconomic reforms usually attributed ineffective implementation to governments' limited management capacity. Economists in international organizations claimed that in some Latin American and African countries, adjustment policies were not effective because governments deviated from recommended prescriptions or were not able to carry them out in proper sequence (Corbo and de Melo 1987). But Streeten concluded from his (1987) review of economic reform policies that no single set of prescriptions – neither *laissez-faire* market policies nor central bureaucratic control – guaranteed success. Governments that were successful in implementing structural adjustment policies:

> knew in which areas to intervene and which to leave alone, and how to conduct the interventions efficiently. The failures illustrate not only excessive government intervention, but also unwise and inefficient intervention in some areas, and inadequate intervention in others.
>
> (Streeten 1987: 1478)

Paul (1988) concluded from his study of World Bank sectoral adjustment loans that ineffective implementation could be traced to inadequate institutional analysis by international organizations in the countries where the policies were adopted.

At the core of structural adjustment policies was the assumption that fiscal and monetary policies were the most crucial variables affecting economic growth (Mussa 1987). But in prescribing structural adjustment policies, international financial organizations gave little attention to the political conditions that were necessary to implement complex economic reforms. As a result, widespread political opposition emerged to structural adjustment policies in several Caribbean countries, and austerity measures sparked rioting and popular antipathy in Egypt, Algeria, the Sudan, Jamaica, and Peru (Waterbury 1989). Neo-classical economic analysts in international funding organizations seemed oblivious to the fact that the success of long-term economic adjustment required strong political coalitions and determined political leaders to set the direction for development. Strong political leaders in Singapore, Taiwan, and South Korea, for example, were able to sustain their regimes and to guide development through a combination of political controls and economic incentives to build supporting political coalitions (Rondinelli and Montgomery 1990).

From detailed studies of Ghana, Zambia, Kenya, Sri Lanka, and Jamaica, Nelson (1984: 102) concluded that the prospects for carrying out stabilization policies were affected by "the strength of commitment to the program on the part of the country's leadership; the government's ability to manage political responses; and the political response that the program evokes from influential groups." In countries where political stability and strong leadership allowed successful development policies to be pursued over long periods of time, the hallmark of success was not slavish devotion to a reform blueprint or even to initial policies, but rather the ability to change priorities and directions as economic, social, and political conditions changed and as earlier stages of development created new requirements or mandated new priorities (Rondinelli and Montgomery 1990).

Although political stability and strong leadership seemed to be important characteristics of countries that successfully adjusted their economic policies, the longevity of charismatic political leaders lacking a development orientation neither guaranteed political stability nor assured the creation of effective guidance systems for responding to economic and social changes. The long reign of Ferdinand Marcos in the Philippines, for example, undermined the production system and weakened guidance institutions (Overholt 1986).

More important was the ability of strong leaders to use long periods of political stability to create more permanent and responsive institutions for development administration. The strong emphasis on macroeconomic reforms in structural adjustment policies often overshadowed the need to attend to other policy issues that were essential to development. Most developing countries that had been successful at adjusting their economies over the years experimented with different combinations of import substitution and export production policies at different times, and adjusted domestic economic structures to changing international market conditions. Reflecting on differences in developing countries" ability to adjust to changing economic conditions, Krueger (1987) pointed out that governments in countries like Korea did not succeed by formulating policies that were always correct. Success was due to their leaders" ability quickly to recognize and correct mistakes when events overcame policy decisions, and to mobilize support for policy changes.

However, not all developing countries in which the IMF and World Bank prescribed structural adjustment policies had the political leadership to implement the policies successfully. For example, the slow progress toward economic and social changes that could improve living conditions in many African countries may be explained by the inability of leaders and institutions to cope with the complexities of macroeconomic adjustment. Adamolekun correctly observed that:

> what is particularly striking about the nation-building phase is how the complexity of the problems, in contrast to the undisputed single goal of national independence under the nationalist struggle phase, has confounded and humbled virtually all post-independence leaders.
>
> (Adamolekun 1988: 101)

After a decade of structural adjustment experience, the World Bank (1990) reported in 1990 that, although many developing countries were better off economically, the number of people living in poverty – that is, struggling to live on income equivalent to less than $370 a year – had risen to more than 1.1 billion. Of these, about 630 million – almost 20 percent of the population of developing countries – were living in extreme poverty with less than $275 in annual consumption. More than 110 million children in developing countries still lacked access to primary education. Substantial portions of the population in South Asia, Sub-Saharan Africa and Central America could not obtain access to: employment; sufficient income to rise above the subsistence level; or basic needs.

The changes in international economic conditions and development assistance strategies that occurred in the 1980s and early 1990s were persistent reminders of the complexity and uncertainty of development and of the difficulties of planning and managing development programs and projects effectively.

4

DESIGNING DEVELOPMENT PROJECTS

The limits of rationalistic planning and management

Ironically, the planning and management procedures adopted by governments and international aid agencies for preparing and implementing development projects have become more detailed and rigid at the same time that development problems have become more complex and uncertain and less amenable to systematic analysis and design.

As the previous chapters explained, international assistance organizations and central planning and finance ministries in developing countries adopted more detailed and rigid planning procedures in the late 1960s and early 1970s in an attempt to anticipate and eliminate many of the problems that had plagued development activities in the past. But in attempting to apply more comprehensive and detailed controls, planners often generated new conflicts and problems. Some arose from the low levels of administrative capacity in developing nations that made it difficult for them to comply with complicated project design and selection procedures. Others evolved from disagreements over the usefulness and efficacy of those requirements. And some were the result of economic, political and social changes that could neither be anticipated nor controlled by development planners and project designers.

Although the structural and sectoral adjustment loans that the World Bank and other international assistance agencies used during the 1980s and early 1990s differed substantially in their characteristics from conventional development projects, they also had to go through many of the same stages of approval. Evaluations indicated that many of these loans encountered similar problems in design, appraisal and implementation as conventional development projects, suggesting that there were inherent difficulties in the way international agencies planned and managed their development activities.

Although there were substantial differences between these instruments of lending, adjustment loans and development projects were not mutually exclusive activities. Indeed, in some countries, World Bank adjustment loans served as "umbrella funding" for a variety of projects aimed at policy or institutional reform. The Bank's energy-sector loan to Pakistan, for example, sought to nationalize and integrate (in a Core Investment Program) projects that had been identified for funding over a three-year period. In Ghana, the Bank used an education-sector adjustment credit to fund reforms in curricula, textbook distribution, teacher training, and school budget management that might otherwise have been undertaken as separate project activities (Nicholas 1988). Even these adjustment loans had to go through processes of identification, selection, approval, and implementation that were similar to the project cycle.

This chapter reviews the requirements and procedures for project planning and design that were adopted by international agencies and governments in many developing countries; explores the reasons why projects so often deviated from preconceived plans during implementation; and examines the factors that limit the usefulness of rationalistic analysis and control-oriented planning and management procedures in development administration.

THE LIMITS OF RATIONALISTIC PLANNING AND SYSTEMATIC MANAGEMENT

A review of experience over the past two decades confirms that – despite the complex formal requirements prescribed for their preparation, analysis, and management – development projects and programs continued to deviate widely from preconceived plans. International organizations, however, continued to attribute delays, cost overruns, changes in objectives, and other deviations to the failure to conform to their prescribed methods – that is, to inadequate design, analysis, and administrative control. For instance, in a performance audit of seventy projects, World Bank (1978: 3) officials found that it was a "fairly common experience for projects to change in the course of implementation." In most cases, the officials reported, "the original design has been proven to be technically faulty or the preparation studies insufficiently detailed to foresee difficulties subsequently encountered."

But international assistance organizations rarely admitted that many of the problems encountered were simply unpredictable, no matter how comprehensively the projects were planned or how much technical analysis was done. Nor have they generally recognized that detailed, rigid, and complex design, analysis, and management procedures may themselves have created many of the problems. Attempts to impose rationalistic and universal standards have, for instance, generated conflicts and tensions among funding agency staff, central government planners, project managers and technicians, and the various groups and organizations affected. Problems also arose from the inflexibility of planning and design procedures – especially when funding agencies attempted to force managers to follow preconceived designs in the face of unanticipated social, economic, and political changes, or when new information about existing conditions threatened the success of a project as it was originally conceived (Rondinelli 1987).

These problems with comprehensive planning, systematic analysis, and central control were not, of course, confined to projects in developing countries. Similar difficulties arose when governments in industrial societies attempted to plan public policies and programs in great detail (Lindblom 1965). Attempts at comprehensive planning for urban renewal, poverty reduction, regional economic development and welfare reform in the United States have often had the effect of making policies less rather than more effective (Rondinelli 1975, Wildavsky 1979). By focusing too heavily on objectives and procedures, planners overestimated the resources available to carry out programs and underestimated the costs of doing detailed and systematic analysis. Comprehensive planning often displaced dynamics of political interaction and eliminated beneficiaries from participatory processes through which different views and perspectives could have evolved.

The description of development policies in Chapters 2 and 3 reviewed the general problems and limitations of the rationalistic planning and management procedures used to implement development strategies. This chapter identifies more precisely the factors undermining the efficient application of those procedures. One needs only a cursory review of evaluations conducted by national governments and international agencies to discover that attempts at comprehensive analysis and control-oriented management generated adverse and often unintended results: costly but ineffective analysis; greater uncertainty and inconsistency; the delegation of important development activities to foreign experts who were not familiar with local conditions; inappropriate interventions by international assistance agency and central government planners; inflexibility; and unnecessary constraints on managers. In addition, serious implementation problems have been created by failure to include intended beneficiaries in the design and implementation of projects and by managers" reluctance to engage in error detection and correction.

Costly and Ineffective Analysis

Detailed and systematic planning is a time-consuming, costly activity that frequently entails long delays in translating policies into action and does not always ensure effective results. In reviewing the efforts of the US Agency for International Development (USAID) to design its projects in the Sahel region of Africa comprehensively, evaluators from the US General Accounting Office (GAO) noted that:

> the project proposals which result are not necessarily either well designed or easily implemented. The lengthy review processes produce advocacy documents which are often too theoretical to be operationally useful. The present design process is complex, requiring between 2 and 4 years for each project.
>
> (US General Accounting Office 1979: 26)

The GAO's evaluation of ten projects in Senegal, Mauritania, and Niger found that more than two years were required to develop and approve project proposals; it took nearly three more months to negotiate them with the governments. In Niger, USAID staff spent three years designing a livestock and range management project, only to find that outside experts and the government considered it too ambitious and that USAID's own review committee thought it too vague. The design of an integrated rural development project in Mauritania required more than three years while planners attempted to collect all of the relevant data and information to design the project systematically. After a series of site visits, however, the agency's review committee decided that the proposal was not feasible and the project was revised again. Thus, after more than three years of intensive planning, it was changed to an experiment to test alternative approaches to rural development, a project that could have been initiated three years earlier and designed incrementally.

The World Bank (1990b) points out that the time required for a project to go through its entire cycle can be as long as ten years. On average, identification requires one to two years, preparation can take as long as three years, appraisal and negotiation require four to eight months. Project implementation averages six years.

The time-consuming procedures used by the United Nations Development Program (UNDP) to plan its projects have been costly to UNDP and have resulted in costly plans for the host governments because they were unrealistic. A UNDP project for developing the master plan for the Lagos metropolitan area in Nigeria illustrates the slow pace at which projects are processed in the United Nations. The initial request for assistance came from the Nigerian government in 1964, but approval for the outline of the project did not come from UNDP until 1973. The project was not completed by UNDP experts and Nigerian counterparts until 1980, and UNDP experts did not depart until 1984 (Onibokun 1990). Over that period of time, economic and political conditions in Nigeria changed drastically. The master plan proposed expenditures of more than 10 billion naira from 1980 to the year 2000, an amount that turned out to be far beyond the capacity of the Nigerian government to raise. As a result, the master plan that took so long to formulate was seriously underfunded (Onibokun 1990). Budgetary allocations for the implementation of the master plan have declined in both real and absolute terms.

Experience with these and other projects attest to the fact that the great detail in which projects are prepared over long periods of time do not ensure successful implementation. The inability of government and international assistance agency planners to forecast the future accurately or to analyze and model all of the variables affecting project implementation often makes their insistence on rationalistic analysis and comprehensive design costly to both the agency and the government.

Inconsistency and Increased Uncertainty

Attempts to do comprehensive planning and design of development projects do not necessarily reduce uncertainty, nor do they make the actions of project managers more consistent with government or international assistance organizations" policies and objectives. Indeed, the long delays that result from attempts at detailed planning can generate even more uncertainty and inconsistency – the very problems that systematic analysis is supposed to overcome. The rapid turnover of personnel in international funding institutions, national ministries and agencies, and among technical consultants hired to assist with analysis and design often leads to increased inconsistency and confusion as the design process drags on.

Evaluations of USAID's projects in the African Sahel (GAO 1979: 27–8) found that "because a new team is usually recruited for each phase of the process, design consistency and efficiency are disrupted." The Senegal Casamance rural development project took two years to plan and had two entirely different design teams. The rural health improvement project in Niger required two years to plan and had three teams working on three different phases of the design. With changes of personnel, conflicts developed among consultants, headquarters staff, and the USAID Mission over concepts and components of these projects. The longer the design process dragged out, the more consensus between USAID field personnel and headquarters staff disintegrated.

Despite the fact that they were planned by formal UNDP procedures, the United Nations" urban assistance projects were often undermined by lack of government commitment and support and by unanticipated changes in policy that were either in conflict with the goals of the projects or made it more difficult to achieve their objectives. In a review of UNDP-funded human settlements projects, de la Barra and Moretti (1988) found that "lack of government support or the lack of flexibility in government official channels either delayed or seriously affected project implementation."

Projects could not be guaranteed of implementation according to plan simply because they were planned by systematic methods. In Liberia, for example, UNDP's assistance for strengthening the national housing authority was counteracted by government policy changes that deprived the authority of its autonomous status. In Argentina, as Hardoy (1990) notes, the government's unwillingness to change housing construction regulations and standards clearly undermined the impact of UNDP assistance in adapting local building materials for low-cost housing construction. Hardoy (1990: 20) also points out that "political changes, particularly changes in ruling parties in provinces and municipalities, gave rise to a great degree of uncertainty regarding technical work and, in some cases, resulted in suspensions and alterations of projects that had been preselected." Frequent changes in personnel – technicians and officials – involved in the project rendered useless much of the training given to former staff members.

The fate of UNDP projects in Nigeria was determined far less by detailed and systematic design than by political and economic instability. "Constant changes of government have led to constant changes in the priority of government and in the level of commitment to project activities," Onibokun (1990: 36) found. This instability adversely affected the implementation of the master plan for Lagos metropolitan area, and changing economic fortunes in Nigeria reduced the government's interest in urban development programs.

Delegation to Experts and Inappropriate Intervention

The complexity of the procedures used to plan and analyze project proposals, together with the scarcity of highly trained technicians in most developing countries, has usually resulted in greater dependence on foreign experts to do the required analyses and to manage development activities. But the delegation of functions to foreign technical experts does not guarantee that the projects will be more effectively implemented or more responsive to the needs of beneficiaries. Indeed, heavy reliance on foreign consultants, who presumably understand and can meet the requirements of the international funding agency, often leads to projects that are unrealistic and inappropriate for local conditions.

The delegation of project design and implementation to foreign experts in Thailand, for instance, has led to heavy expenditures on projects based on incorrect assumptions about local capabilities and constraints, later contributing to serious problems of implementation (Noranitipadungkarn 1977). In other cases inappropriate methods have been applied by technical experts who were not familiar with the culture of the country in which they were working and who designed projects that were not suited to its needs. In a review of UNDP-funded projects for human settlement and urban development, evaluators found that less than 40 percent of the settlements project documents gave any indication that the designers drew on needs assessments or that the projects were based on identification of local technical or financial assistance needs (de la Barra and Moretti 1988). Only half of the project documents mentioned anything about the project's origin and, of these, about half identified the project simply as a follow-on from a previously funded project.

Similar findings emerged from evaluations of UNDP, UNIDO and ILO assistance to small industrial service organizations and for direct support to strengthen small enterprises. Between the late 1960s and 1985, the UNDP funded 642 of these projects costing about $277 million. In addition, UNIDO spent nearly $100 million and the ILO more than $80 million (UNDP 1988). But an evaluation of fifty-six projects funded through United Nations agencies and the Government of the Netherlands" foreign aid program found small enterprise development programs run by small-industry development agencies (SMIDAs) were not very effective because SMIDAs tended to be overcentralized, rigid, and largely urban in focus. They emphasized "hardware inputs" rather than practical advice. Organizations with strong field links, a local presence, and some degree of autonomy seemed to be more effective, but the planners in the international assistance organizations tended to ignore them. These institutional development projects encountered difficulties because planners were not aware of the lack of coordination among organizations involved in rural small-industrial enterprise development. The evaluation indicated that excessive donor intervention created dependency, made it more difficult for host country organizations to manage the programs effectively, and reduced their sustainability (UNDP 1988).

Moreover, rationalistic methods of feasibility analysis, appraisal, and selection have introduced a bias toward the choice of projects that are easy to analyze, but are low priority for development. The insistence of international agencies on using complex feasibility analyses, for example, often leads consultants and governments to propose large-scale, high-technology, capital construction projects because they are considered to be more worthy of the time, effort, manpower, and funds that must be invested in elaborate and detailed analysis than are smaller, more labor-intensive social and human resource development programs. Frequently, the delegation of project planning and design to international consultants resulted in projects that met approved technical and financial analysis requirements, but that were either ineffective in solving local problems or produced adverse consequences (Strachan 1978, Thomas 1974).

Even though they were supposedly subjected to detailed planning and analysis, the design of many UNDP urban development projects was totally unrelated to the conditions under which they would be implemented. One UNDP project promoting low-cost housing construction in Argentina, for example, called for manufacturing brimstone construction panels, even though the production of brimstone and sulphur has been discontinued in the country and the health dangers of sulphur emanation were well known. Evidently, the project was designed and implemented by a foreign expert who had studied brimstone-panel technology and was committed to this approach regardless of the circumstances in the country in which the project would be carried out. The UNDP sent experts in different types of construction technology to Argentina during the design phase of the project without adequately diagnosing the country's needs and conditions (Hardoy 1990).

The procedures of international agencies often produced projects that had to rely heavily on the technical and managerial assistance of foreign technical experts, either because the international agencies did not trust the government to implement them effectively or because government did not have adequately trained professionals. But the delegation of development projects to foreign experts did not necessarily result in more effectively implemented projects or in upgrading the skills of local counterparts. Often conflicts developed between foreign experts and local staff. In many UNDP urban development projects, for example, the UNDP's chief technical advisor (CTA) could not get along with counterparts in government agencies. Although the CTAs" technical qualifications were adequate, their human relations skills were not strong enough to develop and maintain good working relationships with those they were supposed to assist (UNCHS 1987). A review of UNDP urban development projects in Tanzania found that "the inability of the CTA to create harmonious personal relationships with the national staff has hampered project implementation." Kulaba (1990: 40) found that "there are cases where a CTA has operated as if the project was his own and was accused of bringing in things he wanted."

In Nigeria, UNDP selected experts and consultants who, it turned out, were unwilling to live under the somewhat difficult conditions in the country at the time the projects were being implemented. "Housing and irregularity of infrastructural facilities like water and electricity posed a relatively harsh working environment to the consultants and this adversely affected their performance and their level of effectiveness," Onibokun (1990: 37) discovered. "A number of consultants departed prematurely while some of the experts provided for in the contract document could not be sent to Lagos as a result of the prevailing poor housing and infrastructural conditions."

Many UNDP projects failed to strengthen institutional capacity because technical experts simply took over or dominated the work rather than training their counterparts to improve their administrative and technical skills.

In an integrated urban infrastructure development project in Indonesia, supported by UNDP, neither the original design nor subsequent training and operational activities reflected the level of support needed for strengthening the management skills of government officials at the national, provincial and local levels to implement the project. The consultants hired for the project did not have requisite skills in management training and improvement. As a result, in most provinces the UNDP advisors dominated the technical work in the early stages of the project, and in later stages excessive amounts of time and resources were invested in developing guidelines and manuals that tried to impose standardized approaches to feasibility and financial analysis (Lowry 1990).

In Argentina, teams of UNDP experts not only dominated urban development assistance projects, but actually substituted for government officials in performing functions that would normally be performed by public agencies (Hardoy 1990). Because of political turmoil and inadequate numbers of technically trained staff in the Secretariat of Housing and Environmental Management and in other government agencies, UNDP technical experts simply did the work that would normally be performed by government officials. Although this allowed the projects to be implemented according to UNDP plans, it did little to build the capacity of government agencies to deal with urban problems more effectively.

Failure to Involve Intended Beneficiaries in Planning and Management

Many of the planning and management procedures adopted by international agencies and governments of developing countries since the late 1960s were developed by multinational corporations and engineering firms to manage capital construction projects. They conferred an aura of scientific precision that encouraged administrators in international assistance organizations to search for quantitative solutions to problems and to rely on technical standards rather than to seek knowledge and insights from those who were supposed to benefit from social or economic development programs. Few of the 200 rural development projects funded by the

United Nations Development Program (UNDP 1979) during the 1970s, for instance, elicited the participation of the people for whom they were intended. In many rural areas, the longstanding and deeply ingrained distrust between rural people and government officials prevented either side from taking the other into its confidence.

The systems management approaches adopted by international assistance organizations during the 1970s reinforced the paternalistic attitudes of central government officials toward the rural poor, and in the minds of these administrators obviated the need to include beneficiaries in project planning and implementation. Only in about one-third of all the UNDP's rural development projects were local involvement or resource contributions equal to those called for in the original proposals. Less than one-quarter of the projects had any effective participation by local residents, and almost none involved the beneficiaries in evaluation. The failure to include beneficiaries in the design and formulation of projects often led to severe management problems later. For instance, projects were assigned to the wrong organizations for implementation, or to one with insufficient administrative and technical capacity to carry it out, or to one so tightly controlled by vested interest groups that it could not serve the intended beneficiaries (UNDP 1979).

The failure even to define who the beneficiaries of projects would be in the initial design caused problems in many UNDP low-cost housing projects. Kulaba (1990: 30) reports that in the UNDP project in Dodoma, Tanzania, the "target group was defined to be the low income earners who made up 80 percent of the population of Dodoma. Such a loose definition created a number of problems during execution." Much larger numbers of people were eligible for low-cost housing than the project could possibly supply and the conventional lending procedures of the Tanzania Housing Bank, which were to provide credit, could not accommodate that many families.

The transfer of inappropriate technology also indicated little knowledge of or sufficient concern about local conditions and needs in planning procedures of international assistance organizations. Evaluators of the UNDP's (1979) rural development projects found that "a major constraint affecting achievement of project objectives is the transfer of technologies without local adaptation." The complex technology that UNDP experts often recommended had little or no advantage over less complex indigenous methods or equipment. Evaluators argued that "even if there is an advantage, it is often nullified by lack of understanding or by resentment of a new idea 'parachuted' into an area without previous consultation with the users."

The procedures for planning IMF and World Bank structural adjustment loans have often been no better in ensuring that potential beneficiaries and essential participants in implementation were included in program design. Claiming that the government of Zambia did not have adequate numbers of trained economists to fashion structural adjustment reforms during the late 1980s, IMF and World Bank economic missions simply supplanted the staff of the finance ministry in conducting studies and analyses, in creating the framework for policy reform, and in guiding the negotiations necessary to enact reform policies (Callaghy 1989). Although they were successful in formulating a reform program acceptable to the IMF, their lack of sensitivity to local political interests created opposition not only among radical political groups opposed to rapid reform but also among the elite whose economic interests were threatened by the reforms. The failure to include in the design of the reforms those groups who stood to gain or lose politically and economically from the policies meant that an opposing coalition was able to form easily but that the reforms had no strong advocacy group within the country. When they were announced, the reforms aroused immediate popular hostility, and they were abandoned by a ruling coalition that had not participated in their formulation and was not committed to their implementation.

Inflexibility and Unnecessary Constraints on Managers

Another obstacle to effective project administration has been the need for standardized operating procedures in bureaucracies that are subject to close political scrutiny. Congressional oversight of the US Agency for International Development (USAID) was so extensive during the 1980s and early 1990s, for example, that the Agency had to report to Congress on any significant change in any major project's design or funding. Federal law contained at least 288 separate reporting requirements. USAID had to provide Congress with more than 700 notifications of project changes a year. The USAID administrator complained that "legislative earmarking and regulatory snarls make even the delivery of A.I.D. funds an increasing problem. The Agency estimated that the level of effort and costs for annual statutory reports to the Congress alone amounts to 140 work years" (USAID 1989: 114).

Although many of the development problems that USAID and other international assistance organizations have to cope with are complex, uncertain, and unprogrammable, their project planning procedures have not encouraged the kind of learning, collaboration and flexibility in design and administration required to meet diverse needs and rapidly changing conditions in developing countries.

When bureaucracies must respond to strong external demands for control, officials emphasize standardization and reliability in their operating procedures. This leads to rigid behavior, or what Thompson (1961) has called "bureaupathology." Under such circumstances, only clearly defensible actions are taken – even when more innovative, creative, and risky approaches may be needed (Merton 1940). USAID, for example, has attempted to defend itself from Congressional criticism by tightening controls over the allocation of funds, procurement, contracting, and project planning, whenever questions were raised about its operations. The regulations tended to be cumulative and were seldom rescinded.

But there is not much evidence to support the contention that highly detailed, preconceived designs and centrally directed management systems make projects easier to implement or control. Indeed, the difficulty of managing development projects according to detailed preconceived designs was pointed up in the US General Accounting Office's review of USAID's Sahel projects:

> The magnitude of the AID project managers" tasks is enormous. They are to manage the transfer of technical assistance – new knowledge and technology – to the Sahel people. Yet many of the Sahel people are poorly educated and are oriented to tribal customs not all AID development programmers fully understand.
> (GAO: 1979: 5)

The conditions under which the projects had to be implemented were often so constraining that field managers could not adhere to the blueprint plans; nor could they quickly change these conditions to make them compatible with the plans.

A good deal of evidence from evaluations of USAID operations suggests that the control-oriented management systems used in the Agency did not, in fact, give its administrators or the US Congress effective control over development projects. A Congressional task force observed that "accountability of US foreign assistance is extensive but ineffective. Accountability is focused on anticipating how assistance will be used rather than on how effectively it is and has been used" (US Congress 1989: 29).

In reality, experience suggests that the most valuable managerial skill in implementing development projects is not the ability to conform to preconceived plans or schedules, but the ability to innovate, experiment, modify, improvise, and lead – talents that are often discouraged or suppressed by rigid designs and centrally controlled management procedures. What leads to success is the ability of managers to design and manage simultaneously; to test new ideas and methods continuously no matter what the circumstances in which they find themselves. This managerial flexibility, however, is often squashed by officials in the headquarters of international agencies or national ministries who insist on conformance to detailed plans and rigid management procedures. In such situations, the major criterion of success for many project managers is their ability to conform to plans or programs designed in aid agency or national ministry headquarters, rather than their ability to seize local opportunities in order to achieve a project's purposes or to modify goals to reflect changing or unanticipated conditions.

Reluctance to Engage in Evaluation and Error Detection

Finally, the complex, rationalistic, and control-oriented systems of analysis and planning in international assistance organizations have often constrained development administrators from discovering when and how projects stray from their designed paths during implementation. The emphasis on meeting schedules and achieving preconceived objectives makes them reluctant to uncover and correct mistakes. Often, sponsoring agencies or funding institutions ignore the results of monitoring and evaluation. Thus, they do not know when projects deviate from plans or the ultimate effects of those deviations on beneficiaries. Few projects sponsored by international assistance agencies have included provisions for collecting simple baseline data prior to the start of the project and for collecting the information needed to measure progress during its execution (APHA 1977). The UNDP (1979: 27) found that in most of its rural development projects, "baseline data were usually missing, and in those agricultural projects where outputs were clearly specified, most were related to increased aggregate production without assessing distributional effects." As a result, the success of many projects was measured by the amounts of resources expended or inputs used during implementation, rather than by the quality and quantity of outputs, the impact of results on beneficiaries, or the nature of changes attributable to a project's successful operation.

Although the UNDP, World Bank, USAID, and other international organizations as well as many governments of developing countries have elaborate formal project evaluation procedures for monitoring and reviewing ongoing projects, evaluation is less rigorous and accurate than headquarters procedures imply because of the lack of baseline data and the inherent difficulties of measuring actual results and impacts on intended beneficiaries. Kilby (1979: 309) has pointed out in a review of UNDP projects that "it is usually in the interest of all parties to conclude that, save for the most visible failures, every project attains an acceptable level of success." The desire of administrators to place their projects in the most favorable light and the unwillingness of international organizations to embarrass recipient governments, has also inhibited objective and accurate evaluation. An assessment of World Bank public administration reform projects carried out between 1978 and 1988 through fifty-nine structural adjustment loans and forty-three technical assistance loans to developing countries found that the World Bank gave implementation of the reforms little attention through monitoring, supervision, or evaluation. "When such assessments do take place, the analysis rarely goes beyond the finding that progress was either satisfactory or unsatisfactory," a World Bank staff member reported (Nunberg 1989: 25). "As a result, documentation about what worked, what failed, and why these results occurred is almost nonexistent."

Thus, the reluctance of international agencies to engage in error detection and correction makes it difficult for anyone to know when projects are deviating from their planned paths; it reduces the ability of managers to learn from past mistakes and limits their capacity to redesign projects when they meet obstacles or difficulties, or when they are not responding to the needs of beneficiaries.

CONSTRAINTS ON MAKING PROJECT ADMINISTRATION MORE EFFECTIVE

Some development administrators have argued that the systematic methods of planning, analysis, and management adopted by international agencies and many national governments did not work because they were not seriously tried. But given the dynamic environment of political and social interaction that affects development activities and the complex and uncertain nature of development problems, deviation from plans must be expected. Conditions in developing countries make it highly unlikely that rationalistic analysis, systematic planning and control-oriented management procedures can be made more effective.

In fact, the uncertainty of development problems and the complexity of relationships between developing nations and international assistance organizations make it nearly impossible to plan, analyze, and manage projects in highly rationalistic and systematic ways. The attempt to impose such procedures often creates the kinds of adverse results described earlier. The most serious constraints on making project planning more rationalistic include: difficulties in defining project objectives precisely at the outset; lack of appropriate data and information; inadequate understanding of local social and cultural conditions; weak incentives or controls for guiding the behavior of participants in project implementation; the dynamics of political interaction and intervention; and the developing countries" low levels of administrative capacity to plan and manage in the prescribed ways.

Difficulties of Defining Objectives and Goals Precisely

One of the most frequently cited obstacles to the more effective management of development projects is the imprecision of their goals and objectives. Evaluators of UNDP (1973a) projects, for instance, reported that in many of the technical assistance activities undertaken in Nigeria, objectives were stated so vaguely or imprecisely that funding agency staff, government officials, project managers, consultants and participants were in continual conflict over the most efficient ways of managing them. Another UNDP (1974: 27) study found general confusion "regarding the meaning of long-term and immediate objectives, outputs, and workplans. . . . Similarly, immediate objectives (project effects) were commonly confused with measurable outputs, and sometimes with elements of the workplan which was often in the nature of a reporting schedule." The American Public Health Association (APHA 1977: 11) reviewing 180 health projects in developing countries for USAID, found that "large proportions of the projects appear to be engaged in [the pursuit of] a variety of objectives that are largely implicit and unarticulated."

These findings should not be surprising. In most cases it is difficult or impossible to define goals and objectives precisely at the outset, or to give more than general indications of what can be accomplished when a proposal is initially made – especially for social and human resource development projects in countries with unstable political and economic conditions. USAID (1979: 5) missions have often complained of the difficulty of "tracing out exactly who is affected by an activity and what the long-range consequences are." USAID missions frequently expressed their frustration over their inability to quantify the needs of the poorest groups in developing countries during the 1970s. As the staff of the Philippine mission pointed out, "poverty is an elusive concept. Many definitions and measures have been advanced. All have limitations in methodology and applicability to specific countries" established social values." It is nearly impossible to specify goals for projects aimed at increasing the living standards of the poor when "essential factors of poverty, like dietary habits, housing standards, price differentials and ecological diversity are yet to be explored in any great detail." Specific goals often could not be identified until activities were well under way and the conditions under which they had to be implemented were better known. The best that could be done at the outset of a project was to state objectives generally and to aim at broad targets.

Despite the fact that UNDP procedures required detailed and precise identification of the objectives, goals, and targets of projects they funded, evaluations of UNDP urban development projects frequently bemoaned the fact that in reality projects had vague, unrealistic, and overly ambitious goals and objectives. The projects manifested little understanding of local conditions and needs. They failed to identify the intended beneficiaries, to specify the outputs and intended results, or to include appropriate technologies.

Evaluations described the project design as too ambitious or unrealistic in more than half of the human settlements projects funded through the United Nations Center for Human Settlements (UNCHS 1987). The project documents were often prepared with little knowledge of local circumstances. As a result, the immediate objectives were often too broad and vague to be translated into clear guidelines for action. This was an especially serious problem with the training components. Many UNDP urban development assistance projects undertaken in Indonesia during the 1980s also had unrealistic goals; projects were overambitious in terms of the resources available and the management requirements to achieve them (Lowry 1990).

Despite these recurring criticisms, the insistence of international aid agencies on precise and detailed statements of objectives at the outset in order to facilitate systematic planning, management, and control has often led to game-playing, phony precision and inaccurate reporting, creating severe administrative problems later. In his evaluation of eleven United Nations technical assistance projects for developing small-scale industry and handicraft cooperatives, Kilby (1979: 316–17) found that many of the initial proposals did indeed lack specific goals and targets. But he also noted that the planners and administrators who designed the proposals often unrealistically enlarged the goals and targets that they could identify. Unable to predict precisely what the projects could achieve, but faced with demands for quantitative indicators, administrators overestimated the number of people who would find jobs through project activities by more than five times the actual number who were eventually employed.

Administrators faced with the choice of conforming to requirements for systematic planning and precise identification of objectives and goals (regardless of their capacity to do so), or with losing funding for a project, often provide seemingly precise statements that are inaccurate, unverifiable, or merely window dressing. The objectives are often the ones that planners *think* the funding organizations want rather than those that administrators can achieve. Moreover, projects were often designed by central government officials, international agency staff, or foreign consultants who were not ultimately responsible for managing them, and who were more concerned with formulating a "fundable" proposal than with planning a project that was administratively feasible (Ahmad 1977).

Lack of Appropriate or Adequate Data

Even if all of those involved in development projects wanted to define objectives and outputs precisely at the outset, rationalistic planning and management procedures often require information and data that are simply not available in most developing countries. Such demands have forced administrators to use whatever data were at hand, regardless of their appropriateness or accuracy. Requirements that USAID projects be targeted on the poorest groups in developing countries led some (USAID) missions to complain during the 1970s that the precise information needed to get projects approved and to ensure desired results simply did not exist or was not easy to acquire. The USAID (1979: 7) mission in Pakistan, for example, reported that "the identification of depressed economic groups is beset by problems of serious data scarcity." The USAID mission in Burma pointed out that:

> to a Burnese, *sinye* (to be poor) implies the absence of resources but it also describes a distressed mental state related to material status. . . . The Burmese have never defined poverty in economic terms. With the paucity of data, the international donors have also been loath to undertake this arduous task.
>
> (USAID 1972a: 2)

But collecting more data and information has not by itself overcome the recurring problems of project implementation. Nor has it necessarily made implementation more rational and systematic. Indeed, Chambers and Belshaw (1973) observed in their study of the Special Rural Development Program in Kenya that the tendency to collect all conceivably useful data created confusion about purposes, and inhibited managers from learning while they implemented the projects.

A more critical problem has been the tendency of international organizations to use quantitative methods of analysis that often led project planners to collect only readily available statistical data and to ignore or overlook other, more important, kinds of information. Although they were supposedly based on sophisticated economic analyses, the structural and sectoral adjustment loans that the World Bank made during the 1980s and early 1990s often contained little information about the political and institutional capabilities of developing countries to implement economic reforms. A study of World Bank structural adjustment loans attempting to reform trade and investment policies in twelve developing countries concluded that their effectiveness depended on two sets of factors – the organizational capabilities of public bureaucracies and the political flexibility of governments (Levy 1989). Yet, little information was collected at any stage of the World Bank's project cycle about institutional variables that would affect the implementation of structural adjustment loans (SALs) or sectoral adjustment loans (SECALs). "In general, institutional diagnosis in SECALs focused on the technical aspects of service delivery," Paul (1989: 6) concluded. "However, the impact of interest groups on the working of the institutions, the likely resistance to reforms from the political or bureaucratic fronts and an assessment of risks involved in the proposed reforms seem to have been neglected." Institutional analysis was judged to be weak in 67 percent of the SECALS and 73 percent of the SALs. In another study of World Bank projects containing components aimed at promoting institutional development during the 1980s, Eaves (1989: 46) concluded that the lack or insufficient analysis of institutional factors resulted in "project designs pitched far beyond the capacity of existing institutions."

Inadequate Understanding of Social and Cultural Conditions

Another serious limitation on rationalistic planning and control-oriented management is the difficulty of comprehending at the outset of a project all of the social and cultural nuances that are important for effective implementation. Again, the demand for systematic analysis and management has often led planners and administrators to use information and data that could be easily gathered or manipulated by statistical methods and rarely has this included the social and cultural characteristics of beneficiaries.

In the detailed feasibility studies required by some international organizations, analysts often manipulated vast amounts of unreliable data with complex statistical formulas in order to reach a conclusion on the desirability or undesirability of a project. The more sophisticated the analytical requirements, the more likely that essential but non-quantified information was overlooked. Many agricultural development projects sponsored by USAID during the 1960s and 1970s, for example, were identified from linear programming models, but many of the resulting proposals for assistance to small farmers failed to include information that was vital to the success of the project. It had simply never been collected and quantified. The analyses of projects in Guatemala, for example, did not account for cost or profit conditions of the farmers for whom the projects were intended. Farmers had this information, but it was never solicited from them. The project designers simply assumed that increased agricultural productivity would generate larger profits, an assumption that turned out to be highly questionable. The projects thus did little to improve production or raise the incomes of Guatemalan peasants (Hutchinson *et al.* 1974).

World Bank and IMF structural adjustment loans and economic reform projects in the Sudan during the early 1980s were undermined by the lack of analysis of social, political, and religious variables that would affect the behavior of those whom the reforms were to affect. The government's "Islamicization Program" prohibited interest on all domestic lending and substituted a compulsory alms payment, the Zakat, for almost all direct and some indirect taxes, directly weakening the reforms designed by the international organizations (Brown 1986).

The tendency to design projects without adequate knowledge of local conditions and needs has not been confined to those involving foreign consultants, however. Administrators in central government ministries often know as little as some external consultants about minority ethnic, regional, or religious groups within their own countries. Or they may fail to incorporate what is known into project designs because the information cannot be quantified. An evaluation of eight UNDP settlement and livestock projects in the Sudan, for instance, concluded that all of them had had problems obtaining the acceptance and cooperation of local residents. Those that had been designed to settle nomadic groups ran into special difficulties. "Social acceptance became a particular problem in irrigation projects where nomads were expected to be available for scheduled work in their fields at time periods which apparently clashed with their livestock herding interests," Thimm (1979: 48) discovered. The nomads simply opposed the managers of the projects, even though the pasture conditions in the area were technically suitable to making the program a success. In nearly all cases, the projects had been designed by central government planners who knew little about local conditions and who made few attempts to discover or understand the cultural practices that would be essential to successful implementation.

Weak Incentives or Controls to Guide Behavior

In many cases, international agencies and central government ministries lack the incentives or controls to change or redirect the behavior of project participants. Credit projects for small farmers sponsored by USAID in Latin America during the 1970s, for example, were not very successful because managers could not change the lending practices of local financial institutions, which simply refused to lend to small farmers. Lenders considered the farmers poor credit risks and feared that if they changed their policies they would alienate their large-scale farmer clientele who wielded considerable political influence and were not interested in promoting greater competition or improving the lot of the small farmers. In her evaluation of such projects in Ecuador and Honduras, Tendler (1976: 31–3) noted that there was very little relationship between changes in institutional practices and renewals of USAID assistance, even though "the design of current AID loan programs results in a very low probability that promised changes in the institution's behavior will actually occur, once the loan agreement has been signed." She concluded:

> If AID wants to lend to institutions that are not behaving in the desired way, and expects to bring about changes in those behaviors, it must impose costs on the behaviors it wishes to change. It must make them dysfunctional, instead of functional, to the recipient institutions.
>
> (Tendler 1976: 32)

Even the World Bank, which has explicitly used its lending "leverage" to force changes in economic and institutional practices in developing countries as a condition for approving projects, admits that significant behavioral and institutional changes are difficult to achieve. Institutional reforms have been included in many projects designed by the Bank, but its own staff has concluded that "whereas the physical objectives have generally been achieved, and this can clearly be determined, the institutional objectives are both less frequently attained and their attainment less easily measured" (World Bank 1978: 8).

Dynamics of Political Interaction and Intervention

Among the most severe constraints on making the planning and management of development projects more rational and systematic are the inevitable political influences and conflicts that arise in the process. Despite the scientific and technical aura that rationalistic methods of analysis are supposed to impart, many of the decisions about development projects in international organizations have always been made by political criteria.

The political dimensions of the US foreign aid program have always been reflected in the fact that, although development assistance and economic support funds have gone to more than ninety countries, well over half have been earmarked for less than a dozen countries in which the United States has strategic military or political interests. In 1984, for example, 62 percent of the nearly $4.5 billion allocated for development assistance and economic support went to Egypt, Israel, Sudan, Pakistan, Turkey, Lebanon, Costa Rica, El Salvador and Honduras (USAID 1983). In fiscal year 1988, nearly three-fourths of economic security and development assistance was earmarked for Israel, Egypt, Pakistan, and four Central American countries (USAID 1987).

The World Bank, which publicly insists that its allocations of resources are based on sound and objective investment principles and on technical analyses of investment proposals, is frequently influenced by political pressures of member governments (Ayres 1984). The United States government, as a principal contributor to the Bank and a powerful member of its governing board, regularly succeeds in influencing Bank policies and loan decisions. During the Reagan administration, heavy pressures were placed on the Bank to shift from a basic human needs to a macroeconomic growth strategy.

Both the Reagan and Bush administrations also placed direct pressure on USAID to move toward a private-sector development strategy during the late 1980s and early 1990s (McGuire and Ruttan 1990). During the 1980s, the Reagan Administration forced USAID to become "more responsive to the needs of US foreign policy than to the needs of developing countries," McGuire and Ruttan (1990: 154) observed. "A major preoccupation of the administration was using USAID programming effectively in order to achieve foreign policy objectives." Reagan appointed a former political campaign advisor as administrator of USAID and allowed an influential conservative political lobbying group to monitor the internal activities of USAID during his term of office.

But the rationalistic processes of planning and management in international organizations has usually allowed them to mask the overtly political influences on their decisions. These processes often encouraged planners to ignore or suppress political differences when projects were being considered for funding, only to see them erupt in conflicts that undermined implementation. Sometimes after proposals had been submitted and approved by funding agencies and recipient governments, one side or the other became disillusioned when the details or management implications emerged. An audit of a USAID-sponsored insect-control agricultural project in Tunisia noted that "there has never been a full meeting of the minds by all interested parties on the purposes of the project and the responsibilities of the respective governments." In the early stages all were agreed on the desirability of ridding Tunisia of fruit flies, but the US Department of Agriculture's technicians were primarily interested in testing techniques potentially applicable in the United States and the western hemisphere. After the details became known, the Tunisian government was so unimpressed with the economic benefits to be derived in its country that it would agree to further participation only if the United States bore the full cost. Another project was cancelled by the Moroccan government because of similar differences in perception about its purposes and methods (USAID 1972a).

Conflicts have also arisen from the politics of project formulation and negotiation. Although international assistance agencies have insisted that projects be appraised and selected on the basis of objective financial and technical criteria, political leaders often intervened to shape a project's scope, components and organizational arrangements.

That has been especially true in the case of large, high-priority development projects, or those in which a government thought that its sovereignty was being impinged upon by international agencies. Although Kenya, for example, had been extremely successful during the 1980s and early 1990s in obtaining IMF and World Bank funding because of its calculated willingness to adopt IMF structural adjustment policies, its relationships with the World Bank were often difficult because the Bank insisted on changes in politically sensitive institutions. The World Bank sought to eliminate the grain marketing board, make changes in import licensing processes, and promote sectoral adjustments that threatened the interests of important political groups. The government therefore sought and obtained concessions in difficult negotiations with the Bank (Lehman 1990).

Thus, despite the elaborate search, identification, and selection procedures employed by most international assistance agencies, evaluations of United Nations assistance projects in Africa and Asia pointed out that "in practice project selection tends to be heavily influenced by considerations which are extraneous to a rational screening process." The two major factors that often sway decisions are "the immediate socio-political events which lead up to the project request and the personal background of the UN advisers assisting in project identification and project formulation" (Kilby 1979: 312).

In his study of UN-sponsored projects in Malawi, Tanzania, and Zambia, Gordenker (1976) noted that presidents and prime ministers in each country intervened directly in negotiations on large projects; sometimes they deliberately generated conflicts with international agency personnel in order to obtain concessions or greater control over design and implementation. The design and funding of the Botswana nickel-copper mines project, for example, involved a long, tortuous process of political bargaining; for more than a decade, the prime minister played an important role in negotiating with the World Bank and a group of private investors to reach a mutual compromise (Ostrander 1974). The final proposal was less a reflection of formal principles of "good management" than a bundle of political compromises minimally acceptable to all parties.

When international organizations refuse to compromise, governments can prevent them from influencing internal policies. After the collapse of several IMF agreements in Zimbabwe, for example, the government simply refused to apply for IMF structural adjustment programs or for World Bank adjustment loans during the early 1990s because of the stringent conditions for trade liberalization the international organizations were trying to impose. Internal political conflicts between white residents and foreign investors who supported adjustment policies and more radical local groups who opposed such measures because of the potential adverse employment and income distribution effects made it difficult for the government to obtain support for IMF agreements. Political pressures from small land owners, peasants, workers and domestic producers kept the government from adopting IMF reform prescriptions, and, when the international organizations refused to compromise, the government simply withdrew from further negotiations.

Although the political influence on project approval has not been well documented, experience with the World Bank's first highway project in Fiji during the 1980s indicated the degree to which political pressure can influence project selection and appraisal. Political leaders in Fiji eagerly sought and were given to understand, first by the UNDP and then by the World Bank, that a project to finance construction of a major highway on the island would be approved even before the feasibility analyses were completed and the project was appraised (Chand 1989). The consulting firm chosen by the government to complete the economic feasibility report was clearly told that its primary mission was to get the project approved. The World Bank appraisal mission, which was supposed to assess the project independently, merely accepted the consulting firm's calculations for user-cost savings, a principal component of the net benefit on which the rate of return was estimated. Thus, the World Bank ended up approving a project that government officials were advocating strongly, without actually making an independent assessment of its financial feasibility. The Bank did impose stringent conditions on the government during negotiations, but later relaxed some of them and eliminated some components of the project during implementation. One estimate of the actual economic return on the project – less than 4 percent – turned out to be far lower than the 16 percent predicted by the consulting firm and the Bank's appraisal mission. The actual return was "less than the opportunity cost of capital in Fiji, but is also less than the borrowing rate," one analyst later discovered. "Clearly then the project turned out to be an economic disaster" (Chand 1989: 252).

Yet, in the face of strong political pressures, there is little that international organizations can do to control projects despite their rationalistic processes of planning and management.

Limited Administrative Capacity

Finally, as noted earlier, attempts at systematic analysis and comprehensive planning have often led planners to "overdesign" projects: that is, to make them too complex or sophisticated for the institutions assigned to implement them. Methods of analysis used by national planners or the staffs of international agencies have often been imposed on indigenous organizations. In USAID-sponsored small-farmer credit and cooperative projects in Ecuador and Honduras, designers required farmers" organizations to use complicated auditing and bookkeeping methods that they did not understand and were unwilling to learn because they did not conform to national accounting practices. The ability of the farmers" organizations to take advantage of programs sponsored by the cooperative federation was therefore severely limited (Tendler 1976).

Governments in the Sahel region of Africa have never been able to absorb and support the extensive technical and financial assistance provided by aid agencies to that poverty-stricken region. The Sahel governments had neither the administrative capacity nor the matching funds needed to make the projects designed by USAID work effectively. Their inability to provide contributions at agreed-upon levels left the projects and the overall development plan for the Sahel in disarray during the 1970s and 1980s. The Sahel governments were not able to recruit managers for many of the projects until two or three years after they had been approved; inordinate delays occurred in providing routine administrative authority to project directors once they had been appointed. Foreign technicians therefore often had no national counterparts until late in their assignments and were thus not able to train them before their departure (GAO 1979).

An assessment of eighteen United Nations Capital Development Fund housing and urban development projects also found that their implementation was weakened because "institutions capable of implementing projects, let alone of designing them, are extremely weak or non-existent. Where such institutions do exist, their meager human, material and organizational resources are insufficient for them to carry out their tasks" (Lemarchands and Niro 1989: 63).

The ability of technical experts and project managers to follow a detailed project design once they were on the job was also extremely limited. In his evaluation of eleven technical assistance projects sponsored by United Nations agencies, Kilby (1979: 313) pointed out that "effective operation of these experts was hampered by under-provision of counterparts, language problems, lack of logistical support from the Ministry and shifting government policy." In some cases, the government altered its requests or demanded changes in the scope or content of a project after it had got under way, because scarce technical and managerial personnel had to be reassigned to other projects or activities on the insistence of international assistance organizations. Moreover, long delays in receiving essential equipment or supplies often slowed down the projects or changed their fundamental nature, requiring unanticipated alterations in activities and outputs.

Projects designed and prepared by headquarters staff of international agencies and by central ministries in developing countries often required high levels of coordination among government agencies and between the public and private sectors in the host countries. Yet the ability of central planning agencies and government ministries to coordinate and control their resources was extremely weak. An evaluation of UNDP (1979: 21) rural development projects pointed out that "coordination between government departments in concepts and actions is enormously complicated by sectoral factionalism." This problem was aggravated by the sectoral specializations within the United Nations system. "The compartmentalization between technical line ministries and their relationships to the ministries of planning and finance finds a virtual mirror image in the UN system of Specialized Agencies and their relationship to the UNDP Resident Representative in the field."

TOWARD A REORIENTATION OF DEVELOPMENT ADMINISTRATION

All of these problems raise serious questions about the ability of international organizations or governments in developing countries to plan and analyze proposals rationalistically and systematically and to design projects in great detail prior to their implementation. The argument being made, however, is *not* that planning and analysis should not be done, or done with great care, but that existing methods, procedures, and requirements that place strong priority on comprehensive planning and design during the preparatory stages of the project cycle are often misplaced, inappropriate or perverse. The complexity of development problems, the variety of factors that must be taken into account and dealt with during implementation, and the inherent uncertainty about the outcome of all development projects suggest that alternative methods of planning and implementation must be employed.

The remaining chapters, therefore, begin to explore such an alternative that is more appropriate to the conditions under which development activities must be carried out. An *adaptive* approach is based on concepts of strategic planning, incremental analysis, experimental design, and successive approximation in decision-making.

5

IMPLEMENTING DEVELOPMENT PROJECTS AS POLICY EXPERIMENTS

Toward adaptive administration

If rationalistic planning and control-oriented management systems are neither effective nor appropriate in coping with the complexity and uncertainty of development problems, what alternatives do planners and administrators have for managing projects more effectively?

One way of coping with uncertainty, complexity and ignorance is to recognize that all development projects are policy experiments, and to plan them incrementally and adaptively by disaggregating problems and formulating responses through processes of decision-making that join learning with action. Adaptive administration allows planners to perform what Wildavsky (1979: 15) calls the basic task of policy analysis: "to create problems that decision-makers are able to handle with the variables under their control and in the time available." This approach sees planning and implementation as the art of creating problems that can be solved through informed experimentation. Courses of action are shaped from lessons of past experience as well as from a more realistic understanding of current and emerging conditions.

This chapter describes a four-stage process of project planning and implementation that can cope with development problems in an experimental, incremental and adaptive fashion. The framework depicted in Figure 5.1 directs the attention of planners and administrators to issues about which little is known or about which there is a great deal of uncertainty. By planning and implementing projects sequentially through experimental, pilot, demonstration, and replication phases, problems can be disaggregated and alternative courses of action can evolve through what Korten (1981: 214–16) describes as three basic stages of learning: (1) learning to be effective in assisting intended beneficiaries to improve their living conditions or to attain other development objectives; (2) learning to be efficient in eliminating ineffective, unnecessary, overly costly, or adverse activities, and identifying methods that are appropriate for larger scale application; and (3) learning to expand the application of effective methods by creating appropriate and responsive organizations to carry out development tasks.

This chapter describes the characteristics of each phase in an adaptive approach to development administration. Illustrations are drawn from actual projects, even though many of those cited were not planned and managed entirely by an adaptive process and did not always complete the four phases from experimentation to replication. The use of a sequential, four-phased process of decision-making does not imply that decisions should always be conceived of as linear, or that projects must always evolve through all four phases. Enough is often known about objectives, probable effects, the conditions under which they will be carried out, and the characteristics of their beneficiaries, so that some projects can skip or place less emphasis on some stages. Nor does it assume that development planners and administrators begin every project with no experience, knowledge or insights whatsoever. Rather it seeks to provide a framework by which planners and administrators can make use of what they already know and to test the appropriateness of that knowledge in coping with new problems under different conditions. The framework is only useful – as is any planning process, management technique or method of analysis – as an aid to judgment. It can assist decision-makers in coping with problems in a more manageable form, to take actions that are grounded in experience, and to adapt them to the conditions and needs of particular groups of people in specific places to arrive at more responsive decisions – the effectiveness of which, inevitably, remain uncertain.

EXPERIMENTAL PROJECTS

Experimental projects are generally small-scale, highly exploratory, risky ventures that do not always provide immediate or direct economic returns or yield quick and visible results. Their benefits are derived from the acquisition of knowledge. They can be useful in understanding development problems, finding more useful ways of coping with basic social needs, assessing a broad range of possible interventions, and exploring the conditions under which

	Project type or stage			
	Experimental	*Pilot*	*Demonstration*	*Replication or production*
Unknowns or design problems	Problem or objective Possible alternative solutions Methods of analysis or implementation Appropriate technology Required inputs or resources Adaptability to local conditions Transferability or replicability Acceptability by local populations Dissemination or delivery systems	Methods of analysis or implementation Appropriate technology Adaptability Transferablity Acceptability Dissemination or delivery systems	Replicability Acceptability Dissemination or delivery systems	Dissemination or delivery systems Large-scale production technology

	Higher		Lower
Characteristics	____	Uncertainty and risk	____________________
	____	Political vulnerability	____________________
	____	Innovativeness	____________________
	____	Addition to existing knowledge	____________________
	____	Need for creative knowledge	____________________
	____	Need for flexibility of organizational structure	____________________
	____	Need for rare or specialized technical skills	____________________

Figure 5.1 A framework for adaptive administration of development projects

development projects must operate. Thus, experimental projects are needed when problems are not well-defined, elements or characteristics of a problem have not been clearly identified, alternative courses of action have not been widely explored and their impacts cannot be easily anticipated. Under these conditions, effective management methods and techniques cannot be confidently prescribed; little is known about the appropriate types, amounts and combinations of resources or about the most effective sequence and timing of interventions.

Substantial evidence from experience with development assistance over the past 40 years indicates that these conditions characterize most development activities. Even structural adjustment loans, which were used during the late 1980s and early 1990s to overcome some of the problems of project lending, were really social experiments. After reviewing the experience of international organizations in getting the government of Madagascar to make sectoral adjustments in agriculture by liberalizing rice markets, Berg (1989: 726–7) concluded that adjustment policies were inherently uncertain activities. The Madagascar experience, he argued, showed clearly the "many-sidedness of liberalization reforms. There is no such thing as a 'simple' process of price liberalization. Nor is there ever a once-and-for-all reform." He found that all of the organizations involved in rice marketing – from the farm gate to wholesale and retail levels – could obstruct the progress of reforms. The impacts and process of reform could not be predicted clearly at the outset, and they had to be adjusted as unanticipated consequences arose during implementation.

But over the past forty years some development programs have been deliberately planned and implemented as experiments, and the lessons of those experiences are instructive. For example, by formulating development projects through a phased, experimental approach, as in South Korea's village development program, *Saemaul Undong*, information could be gathered from a few places, or from a large number of places about a few activities, before embarking on a demonstration or full-fledged national program. During the experimental phases of *Saemaul Undong*, the national government offered all villages a limited amount of building materials with which to launch small, self-help projects. The experience of both successful and less successful communities was analyzed to determine how they organized self-help activities, how they identified and selected leaders, what forms of cooperation they used, and what kinds of government support would be needed to promote self-help projects requiring more extensive cooperation among villagers (Kim and Kim 1977, Kim 1978).

This and similar projects show that such experiments can be used to assess, simultaneously or sequentially, alternative courses of action, and to allocate resources to those that seem most feasible during early phases of testing or observation. Experimental projects can be used to test the applicability of methods and techniques that are transferred from other countries or sectors. If problems arise from unique cultural or ecological conditions, as is often the case in agriculture, population, and manpower development projects, experiments can search for courses of action that are especially well suited to local constraints.

Moreover, experimental projects can lower the risk of innovating. Cuca and Pierce (1977) concluded from their analyses of family planning programs that the costs and threats of failure can be kept small. Different techniques, combinations of inputs and organizational arrangements can be tested on a limited scale or in only a few places, and, if the experiments do not succeed, other approaches can be tried quickly without wasting large amounts of money or unduly embarrassing the government and the sponsoring organization. Small experimental failures may not threaten the survival of a larger program. In addition, experimentation sets in motion a learning process: "the experiments generated new questions that became the subject of subsequent experiments," Cuca and Pierce (1977: 4) found; and "family planning experiments have also contributed to the development of their own methodology."

But experience indicates that complex social projects cannot be designed as conventional scientific or behavioral experiments in which populations are divided into test and control groups, baseline measures are taken, "treatments" are administered, and the differences due to "treatment" are clearly determined. Chambers and Belshaw (1973) observed in the special rural development program (SRDP) in Kenya that it was difficult, if not impossible, to establish valid experimental and control groups in each area. Different ecological, social, economic, and physical conditions made comparison difficult even in relatively homogeneous regions. Moreover, the "before and after" effects of projects may take years to observe, and changes can be attributed to causes other than the experiments. Problems of finding comparable control areas were also reported in the Comilla rural development project in Bangladesh, where the ability to conduct classical scientific experiments was further limited by political pressures and administrative constraints (Choldin 1969). Project staff were required to modify their activities quickly when unanticipated problems arose or when technical problems made preconceived courses of action impossible to follow. Of course, it was not possible to control the behavior of participants sufficiently to determine the experiments" precise impacts.

Because of the constraints on using a scientific approach, the term "experiment" must be defined more loosely. A broad definition was used in the SRDP in Kenya: trying alternative strategies for attaining specific objectives by organizing projects to show if the strategies work and, if so, how well (IDS 1972). SRDP, which compared rural development techniques in different areas, was based on three principles: that it was possible to distinguish among strategies and to attribute results to them; that strategies must have clearly defined objectives, implying the need for criteria by which to assess successes and failures; and, that objectives should be defined in measurable terms so that the extent of their success could be determined.

Defined broadly, there are at least five types of experimental projects:

(1) Those that focus on *problem definition* either make no presupposition about the nature of a problem or assume that all previous descriptions of problems or courses of action are erron- eous or inappropriate. Usually the experiments examine the "symptoms" that indicate dissatisfaction or the existence of a problem. The project is used to diagnose characteristics or aspects of a problem from which remedial courses of action can be identified. For instance, the Puebla project that began the "Green Revolution" in Mexico started with the assumption that little was known about agricultural conditions in tropical areas and that existing cultivation techniques and extension programs were not appropriate. Experimental projects were used to define problems more accurately and to tailor programs to the subsistence agricultural conditions found in different regions of Mexico.

In Colombo, Sri Lanka, the Million Houses Program (MHP) to assist poor slum dwellers upgrade their housing and community services during the late 1980s and early 1990s relied almost entirely on the beneficiaries" own definition of problems. The

National Housing Development Authority (NHDA) assumed that those who lived in the communities knew their own problems better than any outside professionals or technical assistance experts could ever know them and required community development councils (CDCs) to propose the projects they wanted in their neighborhoods (Lankatilleke 1989). The CDCs were elected by groups of 50 to 125 families in a settlement area and helped the families to discuss their needs and negotiate with government authorities to get their projects approved. The CDCs also helped residents to plan the development of their neighborhoods. Once the projects were submitted and the NHDA approved and costed them, it contracted with the neighborhood groups themselves to construct community facilities under the supervision of an NHDA technical officer. Not only were the projects more responsive to the needs of the community, but they generated employment for neighborhood residents and cut back on corruption and incompetence associated with government contracting. The initial experiments were so much more successful than formal government programs of slum renewal and public housing that they were eventually expanded to all urban areas of the island.

(2) Some experimental projects focus on "*unknowns.*" They seek solutions to problems that are only partially defined. The knowledge gained through experimental activities is used to refine definitions or to distinguish among elements of the problem in different areas.

The experience of the Pakistan Academy for Rural Development with the Daudzai project – which sought to provide basic social and health services, upgrade human skills, and integrate service delivery in rural areas of Peshawar – also confirmed the value of experimental and pilot projects for these purposes. Officials noted that the "Academy learned very important lessons from this experience," the most important of them being that rural people were "in the best position to identify their own needs because they had been living in the villages for generations and no one else could be sensitive and alive to their needs" (Khan 1978: 142). Planners found that rural people were often better able to suggest solutions to local problems than were local government officials. The plans that emerged from experimental and participatory activities were far different than those designed previously by government officials without popular participation. Surveys and reconnaissance studies made planners realize that the needs of villages varied widely and that unless the process was experimental, probing, and participatory it would be impossible to plan appropriate or responsive projects for village development (Khan 1978).

(3) Other experiments seek the most *effective means* of attaining objectives that are already well defined. Goals may be stated broadly and the problem is to determine the most appropriate means of achieving them. Planners of the SRDP in Kenya, for example, had determined that rural poverty in that country was due to low levels of income, poor standards of living and the lack of self-sustaining economic activities. Thus, the immediate objective of SRDP was to fill productivity and equity gaps. The projects explored alternative ways of increasing marketable agricultural output, increasing wage employment, improving public services and decentralizing the planning and administration of government programs. Some experiments sought to increase output through agricultural, commercial, or industrial enterprises, and reduce unemployment through public works projects or by creating private enterprises. Others tried new ways to improve public services through extension, training, social, health, and educational programs, and increase popular participation in decision-making at the district and local government levels. Those alternatives found useful in the experimental phases became prototypes for pilot and demonstration projects.

During the 1980s and early 1990s, the Grameen Bank project in Bangladesh was successful in providing credit to more than 400,000 borrowers, 82 percent of whom were women, who could not get access to loans from the commercial banking system. The program resulted from an experimental "action research project" initiated by Professor Mohammad Yunus at Chittagong University. After it was found that loans of small amounts of capital to the poor could expand productive selfemployment without providing technical assistance, the problem was to find an effective means of recovering payments at a sufficient level to sustain the program. Low repayment levels undermined many internationally funded credit projects in developing countries and made commercial banks wary of lending to micro-enterprises and small businesses.

In the early stages, the project's initiators experimented with a village-level, group-borrowing arrangement that relied heavily on peer pressures rather than collateral to assure repayment. An individual with a small business would have to form a group with four others who needed loans for their own separate enterprises. The borrowers needed no collateral and could borrow only small amounts of capital, initially no more than $200. After the group was given a brief orientation about its obligations, two members selected by the group could receive loans. The ability of other members to obtain loans depended on the initial borrowers making payments on time. Loan defaults and missed payments prevented other members of the group from receiving loans until the accounts were current. After the initial period, the group could be expanded to ten or twelve members, and would meet weekly to collect loan payments and share ideas and information for expanding or improving their businesses. Members of the group provided assistance and advice to each other rather than depending on external technical assistance. The experiments showed that Grameen Bank was able to find a more effective means of providing assistance to the poor. By 1990 the Bank was able to make loans to more than 630,000 borrowers and to recover 99 percent of the payments, a record far better than other lending institutions.

Similar results were seen in the Arabari experimental project in West Bengal, India. The project sought new ways of reducing forest depletion – from theft of wood and from encroachment by settlers – and of replanting deforested areas of state reserves (Cernea 1989). The project employed local villagers to replant trees and to take over responsibility for protecting the forests with minimal official interference. At the same time, authorities made sure that adequate fuelwood was available to the villagers at low prices and that they could obtain access to plough-pieces and construction timber from other forest areas. The experiment created a successful approach to forest protection: villagers completely stopped illegal cutting, they earned income from replanting, they reduced the cost to the government of patrolling for poachers, and within five years the forest was completely regenerated. The project was later replicated in other parts of India.

(4) In situations where both problems and objectives are clearly perceived, experimental projects may be used to discover and break "bottlenecks" or to overcome *deficiencies*. Again, the SRDP projects in Kenya serve as an example. Earlier research discovered four major constraints inhibiting rural development:

first, the lack of knowledge, skills and resources available for increasing agricultural production and employment opportunities in non-agricultural sectors; second, inadequate marketing facilities and physical and organizational infrastructure in rural areas; third, ineffective planning and control procedures in local administration; and, finally, inadequate funding for local programs and projects that could help rural people work their way out of poverty. Many of the SRDP projects, therefore, were designed to test methods of breaking these bottlenecks: extension programs, informal training, and formal educational schemes were proposed, and new organizational arrangements were introduced to elicit popular participation and to mobilize resources. New planning and management procedures were tested. Planners experimented with appointing district and area development supervisors, designating "linkmen" in relevant ministries to expedite action on project proposals and to implement programs, introducing new methods of staff deployment, creating local development committees, and setting aside extra funds for local experiments. New equipment was provided to local administrators and they were given opportunities for special training.

Saemaul Undong, Korea's program for raising living standards in rural communities, used a village classification scheme to identify and overcome deficiencies inhibiting development. Planners found that in "underdeveloped villages," basic infrastructure was missing, organizational arrangements for cooperation were weak or non-existent, housing conditions were poor, and few people were active in community affairs. The government encouraged basic infrastructure and farm-to-market road projects in these villages and provided housing improvement assistance and aid in establishing cooperatives (Republic of Korea 1980). All of these projects had to be implemented by the villagers themselves in order to build community spirit and promote local leadership. The government provided guidance, materials and construction standards. In "developing villages," where basic infrastructure, housing, and income were adequate, Saemaul Undong provided assistance in increasing agricultural production. Building bridges and meeting halls, upgrading farm roads, constructing small-scale irrigation facilities and establishing credit unions, would all require stronger cooperation among villagers. The projects had to be organized by local Saemaul leaders who could obtain technical assistance and materials from the government. Finally, in "developed villages," where infrastructure, housing conditions, and services were above average, *Saemaul Undong* concentrated on promoting income-generating and community welfare projects. The program encouraged group farming, off-season vegetable cultivation, construction of common market facilities, and the development of rural factories.

These experiments tested different combinations of activities and allowed villages to proceed from simple to more complex projects only after fundamental deficiencies were overcome and arrangements for cooperation and local resource mobilization were established. This approach allowed Saemaul leaders to identify and solve different types of problems in villages at different levels of development. It started with small projects that focused on overcoming basic deficiencies and that built community spirit and channels of cooperation and went on to expand capacity to take on more diversified and larger-scale projects.

Salmen (1989: 15) attributed the success of the Grameen Bank's program in Bangladesh during the late 1980s and early 1990s to an experimental approach for identifying deficiencies and then formulating its programs to overcome them. "Starting with experimentation in one and then a few villages, learning by listening and observing the people at work and at home, Grameen discovered that community relations were strong and that a major unfulfilled need was credit," he pointed out. By identifying deficiencies and the potential strengths of communities, Grameen could formalize informal commercial bonds into small groups and introduce peer pressure as an effective way of assuring loan repayments.

(5) Finally, since most development programs cannot be designed and carried out as controlled experiments, learning must be derived from "natural experiments." As Choldin (1969) notes in his description of the Commilla Project in Bangladesh, "the natural experiment involves the historical analysis of a situation to understand the dynamics of previous events," and to incorporate the lessons of successful experiments into other projects with similar objectives. Most often, such projects were not designed to be experiments, but turned out to be tests of important methods, approaches or principles of development. Some were spontaneous, but yielded lessons for planned efforts to achieve the same results.

Korten's (1980) studies of the Indian National Dairy Development Board, the Sarvodaya Shramadana Movement in Sri Lanka, Bangladesh's Rural Advancement Committees, the Community-Based Family Planning Services in Thailand, and the National Irri- gation Administration's Communal Irrigation Program in the Philip- pines, were analyses of natural experiments. They not only offered lessons about how to organize projects to be responsive to local needs but provided insights into the characteristics of natural experiments. Korten contends that:

> these five programs were not designed and implemented – rather they emerged out of a learning process in which villagers and program personnel shared their knowledge and resources to create a program which achieved a fit between needs and capacities of the beneficiaries and those of outsiders who were providing assistance. Leadership and team work, rather than blueprints, were the key elements. Often the individuals who emerged as central figures were involved at the very initial stage in this village experience, learning at first hand the nature of the beneficiary needs and what was required to address them effectively.
>
> (Korten 1980: 497)

Administrators were responsive to changes in local conditions, needs, and capacities, and thus the assistance programs emerged from "a learning process in which research and action were integrally linked."

The lessons of these natural experiments also apply to other innovative or exploratory programs. Such experiments usually defy the detailed planning, programming, scheduling, and evaluation required by international assistance organizations. Indeed, the application of control-oriented planning and management techniques may constrain exploration and establish unrealistic expectations that can lead to false impressions of failure. Because they proceed by trial and error, the course of experimental projects is difficult to predict, let alone control. The development of high-yielding seed varieties in the Puebla Project resulted from just such a pragmatic trial-and-error methodology, which fell far short of the requirements of a scientifically designed approach or a well-programmed effort. "We never waited for perfection in varieties or methods but used the best available each year and modified them as further improvements came to hand," admitted Nobel Prize-winning plant geneticist Norman Borlaug (quoted in Poats 1972: 23), whose team discovered the miracle wheat strains. "This simple principle is too often disregarded by scientific perfectionists who spend a lifetime searching for the unattainable in biological perfection and consequently during a lifetime of frustration contribute nothing to increasing food production."

The decision about how long to run experimental projects can be crucial to their success. Problems have arisen in experimental family planning projects that have been either too long or too short in duration. "Short-run experiments may appear successful because the novelty of their approach guarantees a certain level of impact and because they satisfy the demands of a particular market," Cuca and Pierce (1977: 7) have pointed out. "This success, however, may be short run." But long experiments are difficult to sustain politically. The more complex the social and economic variables that affect the area in which experiments are being done, the more difficult it becomes to determine their impacts.

Because experimental projects are often small-scale, they usually require less financial investment than large projects. But they usually need a substantial amount of scarce professional talent. Personnel recruitment and selection is crucial. Experimental projects require people with an ability to provide creative insights as well as technical knowledge. Highly specialized professionals – such as physicians, agronomists, chemists, engineers, or managers – may be needed to deal with technical problems, but there is also a need for managers who can elicit the participation of beneficiaries in designing and operating them. The success of the Comilla Project in Bangladesh was attributed to the ability of staff to learn the most effective means of designing small-scale agricultural and rural development projects from villagers. "The system was developed through a series of trials and errors which involved gathering as much information from the villagers as possible," Choldin (1969: 485) observed. The staff used flexible, pragmatic approaches "to work out methods through interaction with the village clients rather than starting work with preconceived solutions in the form of large-scale plans." A study of small-scale agricultural projects in Latin America and Africa concluded that farmers can play important roles in experimental projects by generating ideas and identifying useful activities, testing the applicability of new methods or techniques and adapting them to local conditions (Morss *et al.* 1975).

Organizational arrangements must also be given careful attention. Although ideas for experimental projects may be identified, financed, and supported by government agencies, they must often be carried out by autonomous or private organizations, such as agricultural experiment stations, industrial research institutes, universities, or voluntary groups with special characteristics. If a project is carried out by an operating agency of the government, the staff may have to be segregated in a project management unit or research and development division, and relieved of routine responsibilities. Cohen (1987: 235) observed that the success in Ethiopia of the Chilalo Agricultural Development Unit (CADU) – an integrated rural development project – was due in part to the fact that the project had been managed quite independently of the rest of the government in its early years, and received technical assistance from Swedish advisors. Later, when the project was incorporated into the Ministry of Agriculture, its managers had to operate in "a more mechanistic and inefficient way." CADU was able to continue providing services without undue bureaucratic constraints only because the Swedish aid agency had also promoted administrative changes in the Ministry of Agriculture that were compatible with CADU's budgetary and management procedures.

There are a number of reasons for segregating the staff of experimental projects, at least in the initial stages of operation. First, they usually need an extraordinary amount of flexibility and time to explore and the freedom to make mistakes, which often are not available in bureaucracies that value routinization, expeditious action, standardization, and conformity. Second, experiments frequently require special equipment, facilities, supplies, and staff with unusual skills and can be severely hampered by inordinate delays in obtaining them. Third, experiments usually require careful attention to a few special activities that would either compete with or disrupt normal administrative tasks. Finally, there is a danger that experimental projects will be assigned to inadequately trained or inappropriate personnel, or that those who are adequately skilled will resent the additional duties unless they are relieved of routine tasks. In the SRDP, for example, junior field staff of some government agencies who were assigned to carry out experimental agricultural development projects complained bitterly (IDS 1972) and often simply ignored the additional work. In other cases, they merely relabeled their regular duties as "SRDP experiments" to avoid taking on additional responsibilities.

Although the managers of experimental projects are often segregated in special organizations, they need the cooperation of other public agencies. But Cohen (1987) found in his review of integrated rural development projects in Africa that cooperation from other organizations is difficult to obtain unless the experimental projects are likely to provide some technical or administrative benefits for them. Thus, the way in which experimental projects are initially described and publicized can have a strong political effect on their future and on the ability of their managers to elicit the support and cooperation of other organizations.

In their early stages; experimental projects need some protection from political pressures which would either terminate them too soon because of mistakes or replicate them too quickly when they show signs of success. Political leaders are often impatient with experiments, preferring direct action and visible results. Unless experimental projects have strong political and administrative support and a sheltered budget or source of funding, they may be prematurely ended if politicians think that they are progressing too slowly or if they yield results that displease government leaders or powerful vested interests. Political pressures to transform them prematurely into pilot, demonstration, or production projects are equally dangerous.

PILOT PROJECTS

Pilot projects can perform a number of important functions: they can test the applicability of innovations in places with conditions similar to those under which experiments were performed; they can test the feasibility and acceptability of innovations in new environments; and they can extend an innovation's range of proven feasibility beyond the experimental stage. They may also serve as small-scale prototypes of larger-scale facilities and test the market for goods and services to be produced by proposed projects. In Mexico, for example, UNDP (1973a) assistance was used to establish a pilot plant to process seaweed and to test the uses of this protein in human nutrition and livestock feed. The pilot project obtained engineering data on commercial production requirements, while the acceptability of seaweed in poultry feeds and human diet was being tested. In other countries, family planning programs have used pilot projects to determine the costs, effectiveness, political acceptability, operational feasibility, and potential impacts of various birth control techniques in different communities (Cuca and Pierce 1977).

Care must be taken during the transformation from experimental to pilot projects not to use too many foreign personnel or too much imported equipment. If this is done, Pyle (1980) concluded from his studies of the Poshak health and nutrition project in India, then local staff may not be able to continue or expand the project on their own. Evaluations (Cuca and Pierce 1977: 7) of World Bank-sponsored family planning programs found that if a pilot scheme is to be replicated, "then it is sensible to limit its resource requirements to those that would be available in the context of the regular program." SRDP projects in Kenya, for example, were modified as they passed from experimental to pilot phases. Financial resources, skilled personnel, foreign experts, supplies, and equipment were gradually reduced, modified, or eliminated during pilot tests until the prototypes used only those resources normally found in the pilot areas.

In other cases, new resources are needed to adapt experimental projects or to adjust pilot schemes during implementation. An evaluation of small-scale agricultural projects in Africa and Latin America found that in adapting integrated rural development activities attention must be given to: (1) the appropriateness of farm size to the proposed technological packages; (2) physical constraints or limitations in the pilot areas; (3) cost of inputs to the farmer and effects on profit conditions; (4) the reliability of service and technology delivery systems in rural areas; (5) the levels of technological complexity in relationship to the farmer's educational background, level of literacy, and skills; (6) labor requirements of the new technologies and labor availability; (7) marketing problems and demand for the goods that would be produced with the new techniques; and (8) the amount of risk to the users in adapting the innovations. Care must be taken especially in evaluating the preconditions for successful adaptation of foreign technology (Morss *et al.* 1975).

Because they must be adapted to many different environments, pilot projects may require more financial investment than experiments. As with experiments, it is useful to provide pilot schemes with a stable and secure source of financing to protect them from the political vagaries of the budgeting process, especially if some initial trials prove unsuccessful.

In pilot projects, the acceptability and usefulness of innovations are paramount issues. The SRDP in Kenya demonstrated that as long as individual farmers retain control over the allocation of their own resources, the factors most directly determining the acceptance of innovations are profitability, costs of adoption, and the risk involved in using them. Where comprehensive, large-scale, or socially disruptive innovations are introduced, administrators and planners must thoroughly prepare the people who will be affected and ensure that they play a meaningful role in the process. In land consolidation and redistribution programs in Latin America, even landholders with unproductively small plots refused to cooperate because they were not informed of the project in advance and were not prepared to participate. Because planners announced the projects after they had been fully developed, "camposinos found it almost impossible to visualize such a complete change in the landscape, their homes, work habits, services and agricultural practices," Nelson (1973: 244) reports. "Many of the concepts were completely unknown to them." Planners feared that if the public had been informed, opposition groups would have delayed, altered, or obstructed the projects.

A substantial amount of evidence from evaluations of rural irrigation projects also documents the importance of community participation in design and management in improving efficiency, effectiveness, and sustainability (Uphoff *et al.* 1986). One of the most successful programs of community participation in communal irrigation was carried out in the Philippines through the National Irrigation Administration (NIA). First in pilot projects and then on a nationwide scale, the NIA began involving communal irrigation associations in every stage of irrigation improvement from initial surveys to financing. After constructing the system NIA conveyed ownership, and responsibilities for managing and maintaining the systems, to local associations (Reyes and Jopillo 1986).

The NIA approach to promoting community management differed substantially from the way irrigation projects had been implemented in most developing countries. In the conventional approach, government agencies focused first on constructing the irrigation system and then on developing a social organization to operate and maintain it. But NIA assisted in organizing or strengthening the irrigation associations before the start of construction (Reyes and Jopillo 1986). Irrigation community organizers (ICOs) worked with farmers to develop and strengthen their association. They prepared the farmers to work with engineers in planning the layout, design, and construction of the irrigation system. NIA required the participation of farmers in the development of the entire irrigation scheme from the design phase to construction.

In comparing twenty-four participatory pilot projects with twenty-two non-participatory projects in five NIA service regions, Reyes and Jopillo (1986) found strong evidence of community management's impact on improving efficiency, effectiveness, and sustainability. Participatory systems allowed farmers to adapt and modify the irrigation canals and structures more frequently to their own needs and thereby made the systems more functional over time. In twice as many non-participatory projects, farmers allowed channels and systems to fall into disuse. Those communities with high rates of participation had a much more decentralized structure, broader representation in their leadership, and more decentralized management of irrigation systems than did non-participatory projects. As a result, participatory projects were able to recruit more farmers, including tillers and landless tenants, into leadership positions in the users" associations. Non-participatory projects had larger representation of local elites and land owners. Because of their broader membership and leadership characteristics the participatory projects had better capabilities for financial administration than non-participatory systems. Finally, communities that took responsibility for managing the systems provided significantly higher equity contributions for construction and paid back a significantly larger proportion of their amortization payments than did non-participatory communities.

The success of pilot projects also frequently depends on the support of strong leaders who are motivated by community spirit. But motivation is unlikely to be sustained unless local leaders also receive more tangible and visible rewards. The success of Korea's *Saemaul Undong* can be attributed largely to the dedication of Saemaul leaders, who were chosen by villagers and who served without pay. They organized and prodded villagers to cooperate in self-help projects and mobilized resources within the community. Their effectiveness was due not only to their own leadership traits, but to their selection by villagers, the training provided by the Saemaul Leaders Training Institute, and the competitive approach used by the government to stimulate village development. Anyone over 20 years old, regardless of educational, income, or social status, who was chosen by his neighbors could become a Saemaul leader. Although these leaders were not paid, a variety of non-monetary rewards were made available to them. One was the prestige attached to local leadership of a program strongly supported by Korea's president Park Chung-Hee, whose personal interest and tireless efforts to recognize and reward effective local leaders sustained the incentive system. Special benefits thus made the burdensome job a sought-after post. Recognition opened the way for some

Saemaul leaders to receive administrative positions in the national bureaucracy. A variety of honors and awards were granted by the president. Saemaul leaders received discount rates for official trips, qualified for special loans for personal businesses or local projects, and could obtain government aid in educating their children. Moreover, because the Saemaul organizational structure was separate from local government, local leaders had access to governors or ministers who could cut red-tape and make quick decisions on local problems or grievances (Kim and Kim 1977). Later, when the political leadership in Korea changed and many privileges were withdrawn, it became more difficult to find motivated people to serve as Saemaul leaders.

Thus, experience with projects and with non-formal education, population planning, community development, and small-scale agricultural development activities (Niehoff 1977) suggests that the following factors must be considered carefully in formulating and implementing pilot projects:

(1) the basic knowledge, information, and wisdom of people concerning their own living conditions, perception of problems, identification of needs, and desirability and practicality of new methods;
(2) the specific and unique ecological characteristics of areas into which innovations will be introduced, not only natural and physical conditions but also their relationships in supporting human and animal life;
(3) an understanding and respect for the diversity of cultural values and norms found within communities, their amenability to change and the degree of control that local people have over the factors that create, maintain, and alter those values;
(4) cultural traits that shape individual behavior and attitudes toward change;
(5) the formal and informal authority relationships within communities into which changes will be introduced;
(6) the leadership patterns and channels of cooperation, participation, interaction, and communication within the community; and
(7) attitudes toward risk, achievement, and motivational incentives.

In the Comilla project, rural institutions were gradually transformed or new ones were created to establish two-way communications between villagers and staff. Educational and training activities were tailored to meet specific and immediate needs (Rahim 1977). The use of local organizations in small-scale agricultural pilot projects in Africa and Latin America provided a channel for farmer participation and for maintaining regular communications, promoting and reinforcing behavioral changes, facilitating the delivery of complementary services, and mobilizing resources for investment in related or supporting projects (Morss *et al.* 1975).

In the pilot stages, project planning must be flexible and responsive. Cuca and Pierce (1977: 77) found that in pilot family planning projects "a more fluid design is an asset since it permits modification in response to environmental changes and freedom to manipulate inputs." But careful attention must be given to such factors as selecting appropriate sites, because little useful information can be obtained from pilot activities located in areas that are substantially different from conditions in other parts of a country where pilot projects will be replicated. Experience with employment promotion projects in Nigeria indicated that careful consideration must also be given to selecting and recruiting appropriate participants – those who are willing and able to give the project a fair test and who are able to make use of successful results (Mueller and Zevering 1969).

Other factors must also be considered in designing pilot projects, not the least of which is adjusting activities during implementation to obtain the support, or avoid the overt opposition, of government officials, political leaders and vested interest groups. Political hostility can doom a pilot project to failure regardless of its potential technical or economic merits. In the Chilalo Agricultural Development Unit (CADU) in Ethiopia, for instance, the inability of project staff to generate political support seriously limited the impact on low income farmers. Cohen (1987) observed that local political elites either resisted or tried to capitalize on CADU. The central government gave the project little of the support it needed to be effective in implementing land reforms, improving local administration, or removing obstacles to local participation. Support came only for those activities that did not threaten the interests of local and central government officials.

Experience with the Ethiopian project also points up another important issue – the need to provide or arrange for complementary and supporting resources. This is especially important in rural areas where the organizations, infrastructure, services, and personnel needed to make a pilot project successful are weak. Thus, the designers of the Intensive Agricultural Districts Program (IADP) in India, who sought to test the most effective ways of increasing food production and income of small-scale cultivators and to broaden the economic base of rural communities, had to provide nearly all of the essential complementary inputs within a single "package scheme." The integrated packages included not only new seed varieties, fertilizers, pesticides, and improved farm implements, but also low-interest credit, price supports, improved marketing structures, irrigation facilities, and public works, as well as intensive farm management training. In India, as in many other developing countries, the attempt to test the effects of any single input without the provision of the others would have had an extremely low probability of success (Brown 1971).

But this finding must be balanced against another: that although a pilot project aims at testing an integrated package of innovations, administrators and planners should do so incrementally. The number of innovations tested at one time should be limited and well focused. Moris (1981) noted in his study of rural development projects in Africa that multiple innovations are difficult to carry out successfully because they place great strains on limited administrative capacity in developing countries.

All of these factors can contribute to the relatively high cost of pilot projects. If they are successful, however, their contribution to productive capacity may offset some of the costs. Staff training and exposure to new methods and techniques may be of substantial benefit even if the tests prove inconclusive. If pilot projects change cultural norms or values, diffuse new technologies or methods, or increase the willingness of people to consider new ideas, they may well be worth their costs.

Like experimental projects, pilot activities must often be designed to protect their staff from undue political interference or pressure to show quick results. Usually pilot projects perform valuable political functions in developing countries by allowing governments to test new ideas or methods under local conditions without committing national leaders to large-scale, uncertain ventures, the failure of which would threaten their prestige and political support. Through pilot projects, innovative techniques, organizational reforms or "foreign methods" may be tested on a small scale, usually without incurring massive resistance or obstruction by those benefiting from the status quo. "Pilot studies do not engage the prestige of the national bureaucracy," Hapgood (1965: 113) has observed. "If one proves unworkable – and it should be stressed that a high proportion of such experiments will probably fail – it can be abandoned or drastically altered without serious loss of face." Pilot, like experimental, projects must often be segregated in semi-autonomous project management units or conducted by autonomous implementation agencies that can provide sufficient political and administrative protection to allow them to run their course.

DEMONSTRATION PROJECTS

The purpose of a demonstration project is to show that new technologies, methods, or programs are better than traditional ones because they increase productivity, lower production costs, raise income or deliver social services more efficiently. Their major objective is to show potential adopters the benefits of employing innovations. Thus, although demonstration projects may evolve from experimental and pilot phases, they must be designed especially to test the adoption of innovation.

Even as the third phase of an experimental and pilot sequence, demonstration projects carry high levels of risk. At this stage, however, the risk is more evenly shared between project sponsors and intended beneficiaries. For example, if new seed varieties, marketing arrangements or cultivation techniques do not produce the results expected by farmers, they can suffer serious financial setbacks and the reputation of extension agents who convinced them to participate can be permanently damaged.

For these reasons, the success of demonstrations depends on a number of principles that Moris (1981: 123) derived from his review of integrated rural development projects in Africa. Effective demonstration projects should: (1) offer low risks for participants; (2) provide visible and substantial benefits at the farm level; (3) offer participants regular access to cash incomes; (4) assist peasant farmers with meeting recurrent costs after the innovation is introduced; (5) avoid expanding welfare services before there is a production base that can yield revenue to pay for them; (6) use innovations that do not depend for their adoption on loan financing in the initial phases; (7) consider the long-term effects of technology transfer because these may be quite different from immediate effects; (8) not be implemented in a way that bypasses local officials, who will remain long after technicians and managers who initiated the project have moved; and (9) build administrative capacity on small and incremental, rather than on large-scale and complex, activities that have a higher probability of succeeding.

Saemaul Undong, for example, was based on the principle that demonstration projects must generate immediate and direct benefits for participants in the form of improved living conditions or higher incomes. Without visible and immediate benefits, participation is likely to be weak, and those who are convinced to participate may do so apathetically or be distracted by other activities. The success of South Korea's village development program was due to the emphasis that leaders placed on practical and needed projects. The Saemaul leader training program demonstrated beneficial results through applied methods. Assessments of the program (Republic of Korea 1980: 90) also underline the fact that Saemaul training programs placed a great deal of importance on the presentation of success stories. They demonstrated what could be done in advanced areas rather than simply advocating abstract principles or ideologies. They inculcated a spirit of diligence, self-help, and co-operation, and emphasized the fact that progress would come through trial and error.

Demonstration projects can also diffuse innovations transferred from other countries. When technology is transferred, the projects should be tested first in a pilot area and then adapted to local conditions and needs. Demonstration projects for small and medium-scale industrial, agricultural, and rural development, as Hapgood (1965) notes, should be profitable, novel, and include all of the practices required to support them. Methods and technologies must be acceptable and appealing to beneficiaries as well as to project staff. They must be compatible with existing cultural conditions and simple enough to be used with a minimum level of education. Knowledge, materials and other resources required for adoption must be readily available or easily created, and relatively inexpensive. Hapgood found, as Moris (1981) did later, that if farmers are to assume a larger share of the risks in agricultural demonstration projects that use transferred technology, they must be protected from losses through direct compensation, crop insurance, subsidies, incentives, or provision of low-cost inputs. Demonstration projects should have a short pay-off period. The new methods should be available to all individuals and groups who are interested in and capable of replicating them.

Demonstration projects must be located carefully and tailored to local conditions. Those geographical areas or populations with the greatest chance of succeeding should be selected first, and others with less favorable characteristics should be chosen later. The experience with integrated agricultural development projects in India indicated that careful site selection was a crucial factor in promoting new farming methods. Among the criteria used in selecting sites were: the existence of a tightly organized and previously successful community development program through which participants could be mobilized and services and inputs delivered to them efficiently; credit, supply and marketing cooperatives through which the projects could be administered; sufficient natural resources and appropriate physical and climatic conditions to promote increased crop yields; the lack of major physical obstacles or land tenure problems that would inhibit or sharply increase the costs of operating the projects; and local leadership willing to try new methods and techniques (Brown 1971).

The organization of the project is also crucial, and procedures for introducing innovations must be carefully devised to overcome initial resistance and to sustain the momentum of early adoption. In nearly every developing country poor people's distrust of public officials makes them suspicious of projects sponsored by national or local government agencies. Moreover, innovations may be resisted because poor people, having lived with existing conditions all of their lives, do not feel the same urgency for change as project staff. The fear of risk is especially important among those living at subsistence levels because the loss of meager assets endangers their survival.

Drawing on experience with agricultural development projects in Africa and Latin America, Kulp (1977) suggested a six-step strategy for introducing demonstration projects:

(1) analyze local conditions to compute the best program for the average farm in the area;
(2) standardize the program as much as possible, consistent with variations in local conditions, so that adoption is easier and management more efficient;
(3) integrate projects into the economy of the area by convincing as many suitable farmers as possible to adapt the program;
(4) saturate the area with promotional activities, even if it means overpromoting the program in the early stages;
(5) concentrate initial projects in a limited number of areas to prevent overextension of resources and manpower; and

(6) accelerate the program rapidly each year after the success of the project has been effectively demonstrated.

Demonstration projects must be gradually and carefully introduced into communities. Experience with community development projects in Asia suggested that a series of well-tested steps should be followed to introduce innovations and that the beneficiaries must be organized to accept them. Haque *et al.* (1977) argued that the role of the initiator – usually project staff members from outside the community – is crucial. The initiator must help to identify appropriate beneficiary groups and work with them in adapting the project to local needs. When projects are aimed at the poorest groups in a community they must be class-biased and deliberately designed to increase the autonomy, self-reliance, and political influence of groups that are usually economically dependent and politically powerless. The beneficiaries must be given sufficient resources to be able to take collective action and "de-linked" from traditional dependencies, especially client–patron relationships. Once demonstration projects have succeeded, linkages must be established with groups in other areas to reinforce change and to allow for lateral transfer of experience. Specially trained personnel with technical training and with experience in the methods being demonstrated, and who are sincerely interested in working with potential beneficiaries, must be available to assist in demonstration projects. There is some danger in assigning this work to the staff of regular administrative agencies until the projects are ready for full-scale replication and a service delivery system has been tested.

Changes that occur in the demonstration area must be monitored, both to measure the impact on beneficiaries and to modify the project during implementation. The results of demonstration projects must be observed for some time after they are completed, since conditions for success may change drastically. Green Revolution technology in Asia, for instance, often showed excellent results during pilot and demonstration phases by raising agricultural yields, the income of small-scale farmers, and the wages of agricultural laborers. But later the results were disappointing. In many places yields decreased, and benefits went mainly to large-scale farmers and wealthy landowners because of changes in the conditions under which the new technologies could be used successfully. Widespread adoption uncovered many technical problems that had not been given sufficient attention in pilot phases. Rising costs of inputs – irrigation equipment, petroleum, fertilizers, and pesticides – made them accessible only to wealthy farmers. The methods were often replicated in areas where conditions were entirely inappropriate (Frankel 1971, Wade 1974). Moreover, initial increases in income owing to higher crop yields were later offset by technological displacement of labor and unemployment. Haque and his associates (1977) suggest a number of criteria for evaluating demonstration projects, including: (1) changes in the economic base of the community and in the distribution of economic benefits; (2) changes in attitudes and behavior of beneficiaries as expressed in their increased self-reliance, solidarity, and collective and creative activities; and (3) changes in the ability of people to initiate and carry out development projects on their own.

If projects demonstrate the effectiveness of new techniques and promote social change, they may be criticized severely by those with vested interests in the status quo. Or, project managers may have difficulty resisting political enthusiasm for premature replication. Careful and incremental adaptation is still important at this stage. Weiss *et al.* point out that the scale of operations is important for demonstration projects:

> because many problems do not become fully apparent until a large-scale operation has been reached. In effect, more of the system is tested in demonstration projects because logistics and support mechanisms, a full range of personnel, and other needs must be met to integrate all of the organizational and physical inputs.
>
> (Weiss *et al.* 1977: 99)

Finally, attention must be given to timely completion of the demonstration project and transition to full-scale production or service delivery. Demonstration projects should be terminated when a critical mass of intended beneficiaries have adopted their methods, techniques or outputs. The sponsoring agency's efforts should then focus on transferring the innovation to operating agencies, or the private sector, assisting with full-scale production or service delivery, and transferring personnel resources and knowledge to those organizations that will carry on the work.

REPLICATION, DISSEMINATION AND SERVICE DELIVERY PROJECTS

The dissemination of tested methods, techniques, or programs through replication, full-scale production, or service delivery projects is the final stage in an experimental series. The major contribution of these projects is to expand productive and administrative capacity. Basic design problems include those of testing full-scale production processes and technology, developing appropriate and effective delivery and distribution systems, transferring production and delivery systems to government agencies or private organizations that can manage them on a broader scale, and maintaining an adaptive and responsive administrative approach after they are transferred. Appropriate scheduling, programming, and coordination mechanisms may be needed to ensure efficient, economical, and reliable production and distribution of goods and services. "In the production stage and even in some of the larger demonstration projects," Weiss *et al.* (1977: 101) note, "an additional need has been the quality of entrepreneurship, of working with and exhorting and coordinating the multiple organizations to achieve production-oriented goals."

Transforming experimental, pilot, and demonstration projects effectively into continuing programs is crucial to sustaining their benefits. In this stage, as in all other phases of development projects, the participation of beneficiaries is a primary factor affecting sustainability. A study of a large sample of USAID projects concluded that "sustainability ratings on projects involving outreach to dispersed participant or client groups are closely related to the degree to which those groups are involved in policy making, planning and program management" (Kean *et al.* 1987: 48). Studies of the sustainability of USAID projects identify a number of factors increasing the likelihood that programs will survive the phasing out of international funding. Among the most important factors are an explicit concern for assuring sustainability in the design of the project. Flexible arrangements must be created through which public or private organizations responsible for taking up project activities can respond to changing circumstances and client needs. Other factors include respect for the social characteristics of groups participating in the project and the strength of institutional linkages among participating organizations (Kean *et al.* 1987). Evaluations indicate that the strength of leadership and human resources in the organizations taking over responsibility for project activities, and appropriate arrangements for raising sufficient financial resources to cover operating and capital replacement costs, also affect sustainability.

Undoubtedly, the most difficult problem at this stage is transferring experimental, pilot, or demonstration projects to government agencies or other more permanent organizations. Pyle (1980: 123) noted correctly that "corpses of pilot projects, particularly in the social sectors, litter the development field." He concluded from his assessment of the Poshak project in India that careful attention must be given in the earliest stages of pilot and demonstration projects to building political commitment and support among those who will decide about their expansion and replication. The base of support must be broadened quickly when positive results first appear and the interest of other government agencies must be attracted to them. "Because pilot projects do not receive much attention, this added support could be very important when it comes time to discuss the scheme's long-term future," Pyle (1980: 143) argues. "Aggressive public relations efforts during the field work are a necessary part of the pilot project's life if eventual adoption is an objective."

Even when pilot and demonstration projects have been recognized as successful, attempts may be made at the end of the experimental phase to absorb them into the bureaucracy and administer them in conventional style. They are also particularly vulnerable at that point to political attacks that could lead to their abandonment or termination. India's highly successful pilot and demonstration projects in community development during the 1950s and 1960s, for example, were never effectively replicated. Sussman (1980: 115) notes that when the time came for a decision about their replication, "the change to a national scale raised problems different from those of a project of limited size and duration." Indian leaders were required to make new, more extensive, and potentially threatening commitments.

> Choices on national priorities involve heavy outlays of capital resources and expenditures. The question that decision-makers may therefore find themselves asking is not, What is the *best* way of doing community development? but, What is the most politically and bureaucratically *feasible* way to do community development?
>
> (Sussman 1980: 121)

In India, political pressures to extend the program's coverage led officials to disregard the well-managed, responsive, adaptive, and innovative approaches that had evolved from the pilot and demonstration phases, and to absorb community development activities into the national extension service, which was organized hierarchically and managed so rigidly that it was nearly impossible to carry them on in an adaptive and responsive manner. From the experience with India's community development program, Sussman cogently concludes:

> that the calculus involved in the development of a national program is complex and depends even more on political and bureaucratic perspectives than it does on what is learned in the field trial, the pilot project or the demonstration project. Thus, we might expect in other cases that the pilot project might well not be adopted as a model for the national program unless its political utility and feasibility can be demonstrated to be the most attractive alternative to decision-makers.
>
> (Sussman 1980: 121)

The greatest problem of transforming experimental, pilot and demonstration projects into larger-scale production or service delivery programs is the assumption that they can be transferred to all parts of a country without further testing or adaptation, and that they can be managed by conventional procedures. But all development projects are somewhat experimental, and even seemingly routine replications of thoroughly tested technology or construction methods often meet unanticipated difficulties when transferred from one cultural setting to another, from specialized project units to government operating agencies or the private sector, or from the demonstration stage to full-scale production.

Organizational problems become paramount, for even when previous experimental activities have resolved technical and administrative uncertainties, new problems tend to arise simply with expansion of scale. Although pilot and demonstration projects in Comilla, Bangladesh, for example, proved to be highly successful under the leadership of the Academy for Rural Development, new operational problems arose when prototypes were transformed into a nationwide Integrated Rural Development Program (IRDP). When the programs were replicated nationwide, many small problems became more serious. One official of the Academy observed that the Ministries of Agriculture, Rural Development and Education, which were responsible for the project, were in constant conflict over the allocation of funds, jurisdiction, and sharing of responsibilities and rewards (Rahim 1977). Lack of financial resources, bureaucratic centralization, and ineffective local administration all plagued the program after it was replicated.

The diffusion of new agricultural technologies has also been severely hampered in developing countries by the difficulty of organizing extension services appropriately and of recruiting and motivating extension agents to disseminate them. Demands made on their time by bureaucratic procedures, paperwork requirements, or the remoteness of headquarters from the farms; the lack of adequate equipment and supplies; and the unwillingness of agents to work closely with peasants and small-scale farmers have been common problems. Agents often came from urban areas and had limited knowledge of agriculture. Their low level of rapport with and respect for their clients created mutual distrust and hostility. Attitudes of superiority and paternalistic behavior quickly reduced agents" effectiveness. Their unwillingness to live in rural areas or to spend sufficient time with farmers made it difficult to gain their client's respect or to learn of their unique problems and needs. Moreover, departments of agriculture often failed to provide the resources needed by farmers to apply new technologies effectively, further alienating the extension agents from their clientele (Trapman 1974, Leonard 1977, Heginbotham 1975).

The choice of organization to carry out production or service delivery projects is crucial. A variety of organizational arrangements can be used, but experience suggests that no single arrangement is universally effective. Each form has its advantages and limitations, depending on the administrative, cultural and political conditions under which the projects must be implemented. Among the organizational alternatives currently used to implement projects are (Rondinelli 1979a):

(1) an existing government department or ministry, in which the project is implemented as part of an ongoing program without creating a distinct project management unit;
(2) a distinct project management unit, within an existing government ministry, that is given all of the resources needed to implement the project or provided with functional support from specialized departments in a matrix arrangement;
(3) an autonomous implementation unit outside of the regular government operating structure, with sufficient resources and authority to implement the project, independent sources of revenue, recruiting, hiring, and training capability, the ability to

pay higher salaries and provide greater amenities than regular civil service agencies, and the authority to contract for foreign technical and financial assistance;

(4) a decentralized field unit reporting to a central government agency, usually created to undertake functionally specialized or regional development projects that cannot be implemented directly by a central government ministry;

(5) an interagency coordinating committee, which attempts to integrate the resources of a variety of ministries, subordinate units of government and private organizations or groups, with staff seconded from one or more of the ministries for temporary duty;

(6) a private contractor or production firm that undertakes construction and operation of the project under government supervision, or on a "turnkey" basis, whereby a private firm constructs the project and then turns it over to a government agency for operation;

(7) a lower level of government through devolution of project management functions to provincial, state, or local units, with or without central government supervision and monitoring; and

(8) joint ventures in which government and private firms share responsibility for the construction, operation and delivery of goods and services, and maintenance of the project, with duties, powers, and responsibilities of each party clearly delineated.

With the increasing financial pressures that followed from the worldwide economic recession of the early 1980s and the fiscal austerity policies adopted in the late 1980s and early 1990s, governments in many developing countries have been seeking ways of including non-government organizations and the private sector in service provision (Rondinelli and Kasarda 1991). Governments in some countries are transferring service delivery functions to non-government organizations such as cooperatives, trade unions, women's and youth clubs, and religious and ethnic groups (Ralston *et al.* 1981, Cointreau 1982). In Latin American countries, many public functions, such as vocational education, are being transferred to private organizations (Rondinelli *et al.* 1990). In Argentina, for example, the government actively involves banks, trade unions, and private companies in forming and financing "associated schools" to provide specialized training in urban areas (Cowen and McLean 1984). Among other methods used by governments to disseminate innovations are public–private partnerships. These include: joint ventures, in which public and private organizations formally or informally work together to implement urban development activities; joint investment, in which public and private organizations finance facilities and infrastructure or urban development projects; and turnkey projects, in which governments agree to buy or lease completed facilities constructed by private investors, or vice versa.

The World Bank has supported public–private partnerships for urban housing and service provision through sites-and-services programs. Beginning in the 1970s, many governments in developing countries sought alternatives to public housing to meet growing needs for low-cost shelter. Through sites-and-services programs, government housing agencies assemble, clear, and provide with services and infrastructure land that is divided into home sites and sold to the poor. Poor families contract with small-scale firms or private builders to construct their dwellings, or they build the houses themselves, usually with subsidized materials or with credit provided at low interest rates (Rondinelli 1990). Central agencies or local governments are also contracting with private organizations to help provide services that public agencies cannot offer efficiently or effectively. Contracting for services allows governments to arrange with private organizations to provide services, facilities, or infrastructure that meet government specifications for quantity and quality (Savas 1982, Ferris and Graddy 1986).

Governments in developing countries are also contracting with the private sector as a way of sustaining services that are funded in part by international assistance organizations. In Latin America, the governments of Chile and Guatemala offer territorial concessions in large cities for long periods of time to companies that procure, purify, distribute, meter, and charge for water. In both countries, tariffs are approved by the national government, which also monitors water quality. In Peru, the government contracts out to private-sector organizations specific activities involved in water supply, such as meter reading, computer services, and billing and collection. Private firms are also supplementing or extending local government services. In the Central Region of the Eastern Sudan, for example, the Regional Ministry of Health contracts with private organizations employing sweepers with donkey-carts to collect dry refuse and garbage house-to-house in small cities (Lewis and Miller 1986).

In other countries, governments are providing guarantees or fiscal incentives to induce private organizations to provide services that contribute to urban development. The government of

Barbados, for example, created a Housing Credit Fund (HCF) in the Ministry of Housing and Lands through a loan from USAID. HCF provides capital at below market interest rates to private banks, trust companies, the Barbados Mortgage Finance Company, and other financial institutions to make loans – using regular commercial procedures – for low-cost housing in urban areas. The HCF is a revolving fund that has expanded substantially the role of private commercial lenders in extending credit for housing to low-income households. Moreover, the HCF works with private builders and local officials who are responsible for building and land-use regulations, to plan and obtain approval for the construction of housing units that low-income families can afford (LaNier *et al.* 1986). In those developing countries where there is a strong private sector, private enterprises can and do play important roles in providing services and facilities that are, in reality, quasi-private goods. The greatest opportunities for privatization are in services such as transportation, water provision, health care, and trash collection, for which users can be identified, the spillover effects are low, costs are divisible, and consumers are able and willing to pay for service improvements.

Private organizations are also finding opportunities to provide services such as education and health care at a higher level of quality than those provided by government. The private sector often provides specialized health care in urban areas more effectively than public institutions. The private sector and non-government organizations will also increasingly provide services such as utilities and energy generation that require technical specialization or technological sophistication (Rondinelli and Kasarda 1991).

Finally, private enterprises, especially those in the informal sector, are playing a larger "gap-filling" role: that is, they provide more services for poor households that are unserved by local governments. The informal sector usually extends services or goods that can be offered at very low cost or in small lots to cater to the needs of the poor. Water vendors, for example, are common in cities where municipal water systems have not been extended to slum communities. Non-governmental organizations can also assist in extending public services through self-help or public–private partnership programs. There is also substantial scope in cities for public and private organizations to work together more effectively through contracting and turnkey projects (Rondinelli and Kasarda 1991).

Regardless of the type of organization used to implement development activities, Moris (1981: 24) found that certain principles are essential for adaptive and responsive administration. He observes that:

(1) the success of a project depends ultimately on finding managers who are dedicated and committed to its goals and purposes and who are given discretion in making choices as the need arises during implementation;
(2) supervision procedures should be kept simple and the chain of command kept short;
(3) the project should be kept under the control of a single organization, but contacts should be maintained with others that can support and promote its activities;
(4) staff should be recruited from among qualified people who have served in the area where the project will be located;
(5) staff should work through local officials when contacting local residents, and those who work directly with beneficiaries should be well trained, supervised and supported;
(6) staff, consultants, and contractors should be chosen on the basis of their past performance;
(7) political constraints should be taken seriously and, when it is possible, the priorities of politicians who can affect the success of the project should be accommodated;
(8) resources and attention should be focused on only one or two major activities at a time, starting small and aiming at complete success with each step of expansion; and
(9) subordinates should be given as much experience as possible and be allowed occasionally to stand in for experienced leaders in order to create a pool of future leaders who can carry on the project's activities.

Although it may be necessary to standardize and institutionalize many tasks during replication and dissemination, care must be taken to maintain an adaptive and responsive approach to administration. Standardization is possible, but often risky.

Where professionals are scarce or replication depends on high levels of participation, it may be desirable to use local paraprofessionals or to devolve administrative responsibilities directly to local groups or private voluntary organizations.

The changes required of bureaucracies or beneficiaries, however, cannot be so foreign to their customs and traditions that they inhibit acceptance or impose practices that obstruct adaptive administration. International assistance organizations especially should devote more attention to identifying and using informal methods of decision-making and interaction. Coordination and co-operation in most developing nations are achieved not by command but through informal, personal networks of interaction. They usually depend on complex personal obligation and exchange relationships that cannot, and probably should not, be displaced quickly with formal administrative or organizational mechanisms if they can be made to work effectively. Grindle (1977: 40) noted that in Mexico, for example, bureaucratic "informal exchange networks develop because they are perceived to be, and are in fact, an efficient and effective means of goal attainment." These informal networks of personal exchange and obligation extend throughout the bureaucracy:

> In Mexico, exchange relationships bind public officials from various institutions together for the pursuit of policy goals: they serve to connect individuals within one agency for defense against the functional encroachments of another; they tie the bureaucratic elite to the political chiefs and make possible intra-governmental problem solving; and the manner in which they link the nationally oriented regional elite to the bureaucratic center is useful in understanding problems of policy implementation.
>
> (Grindle 1977: 40)

Although more detailed programming is possible after pilot and demonstration stages, overemphasis on mechanical efficiency and heavy reliance on detailed programming, scheduling, and networking techniques may lead administrators to overlook broader political issues. And although political threats to the survival of projects may be less severe in this stage than in experimental, pilot and demonstration phases, replication and production projects face potential problems arising from political apathy and administrative inefficiency that can seriously obstruct their implementation.

Experience also suggests that development administrators not only have to possess appropriate technical and managerial skills, but must also provide strong leadership, especially in mobilizing resources, interacting politically with other leaders, motivating staff, clients and sponsors, building a network of organizational support, and solving problems creatively.

Without strong administrative and political support, conventional methods of planning and management are inadequate to facilitate project implementation or promote the adaptive and responsive approaches to administration that have been described in this chapter.

6

REORIENTING DEVELOPMENT ADMINISTRATION

Principles, problems and opportunities

Applying the principles of adaptive administration requires changing the perceptions of planners and administrators about the nature of development and of social problem-solving. Conventional approaches to development administration are based on inaccurate assumptions about the process of development, the tasks of development planning, the conditions under which change is possible, and the means through which it occurs in developing societies.

In their review of *How the West Grew Rich*, Rosenberg and Birdzell (1986: 33) point out that the industrialized countries succeeded in achieving steady economic progress because they developed a pervasive institutional capacity and the human resources to innovate and adapt to change. The creation of this institutional capacity and human resource base depended in turn on the diffusion of authority, the widespread use of experimentation to find appropriate technology and effective means of production and marketing, and the emergence of diverse ways of organizing economic activities. In addition, western countries developed because many organizations could provide "incentives which combined ample rewards for success, defined as the widespread economic use of the results of experiment, with a risk of severe penalties for failing to experiment." Progress in economic development was thus based on three factors: autonomy, experimentation, and diversity.

In contrast, the approaches to development administration prescribed by international assistance organizations and macroeconomic theorists for developing countries during the 1950s and 1960s were reflected in two variations of a strategy for improving central governments" control and direction of development. First, a "tool-oriented" approach used by many international assistance organizations assumed that development could be accelerated merely by transferring administrative procedures and techniques from industrial countries, especially from the United States, Britain, and France, to governments in developing countries (Siffin 1977). Second, the political modernization approach assumed that political processes and administrative structures had to be thoroughly reformed to allow governments to plan and direct economic development more efficiently. Both approaches assumed that the central government was the most crucial institution in the development process and that improvements in its ability to plan and manage the economy would lead to rapid economic and social progress.

Those who believed that the government's administrative capacity could be expanded by transferring procedures and techniques from industrial nations sought to establish within developing countries administrative procedures that were "rational" and politically impartial. Advocates of this approach insisted that development administration was concerned with the "technical procedures and organizational arrangements by which a government achieves movement toward development goals" (Katz 1970: 120). They were concerned with rationalizing the methods used by governments to formulate and implement national plans and development policies (Riggs 1970). This approach prescribed a hierarchical organization of bureaucracies in the Weberian tradition, creation of politically impartial civil services in the British tradition, the adoption of western personnel administration systems, the transfer of many service functions to public enterprises, and introduction of program budgeting and rationalistic analysis.

Political modernizers, on the other hand, believed that the transfer of administrative procedures and techniques from western democracies was necessary but not sufficient. They viewed development administration as "social engineering," and national governments as the prime movers of social change. In this tradition, Landau (1970: 75) defined development administration as a "directive and directional process which is intended to make things happen in a certain way over intervals of time." Others perceived of it as a means of improving the capacities of governments to deal with problems created by modernization and change (Lee 1970, Spengler 1963). Development administration would be the instrument for transforming traditional societies. Therefore, unless the entire political system was reformed and modernized, governments in developing countries could not adequately direct and control social and economic progress.

These traditional approaches to development administration came under heavy criticism during the 1970s. Siffin (1977) concisely summarized the weaknesses of the "tool oriented" approach: in attempting to create rational, politically impartial, and efficient government it assumed the existence of western values that usually did not prevail in other societies. Its advocates assumed that complex social problems could be solved with modern administrative techniques. But in many countries the transfer of western methods simply introduced predetermined solutions and inhibited the experimentation and innovation that would enable a wide variety of organizations in developing countries to deal with unique problems as they arose. Moreover, the tools were transferred from well-structured institutions in industrial societies to loosely organized governments in the developing world, where the tools often could not be used effectively. Indeed, many procedures and techniques, such as program budgeting and systems analysis, were transferred before their effectiveness had been proven in the countries where they originated. In countries where the techniques actually took hold they sometimes created powerful technocratic classes that were out of touch with the problems and needs of their own people and especially with those of the poor. Finally, the tools of western administration were concerned primarily with maintenance functions and did little to improve either the government's or other organizations" capacity to innovate and experiment.

The Weberian model of organization was especially inappropriate for developing countries because it overlooked or ignored the high level of uncertainty attending the implementation of development policies. It was primarily modeled on European government systems in which routine and standardized administrative procedures were suited to solving marginal and easily identified problems. But in developing countries the only certainty is that the course of development is uncertain. Solving one aspect of a problem or one set of problems merely uncovers or creates new ones, many of which cannot be anticipated through rationalistic analysis and comprehensive planning and cannot be dealt with through standard operating procedures. The difficulties that international agencies and central governments faced in formulating policies and projects comprehensively, and managing them systematically, arose from the fact that many hidden obstacles and problems could not be identified in advance and therefore could not all be dealt with in the initial design.

Similar criticisms were made of political modernization theories. They were ethnocentric and based on philosophies and values that often rendered them useless or perverse in developing nations. Pye (1965) points out that modernization theory never yielded a concise definition of political development. The concept was variously defined as creating political prerequisites for industrialization, creating government institutions with characteristics similar to those found in European states, reforming legal and administrative structures in the American or British tradition, mass mobilization and participation in political processes, creating procedures for orderly political succession, or democratic sharing of power and authority.

The structural adjustment and macroeconomic reform policies of the 1980s and early 1990s tried to swing the pendulum too far in the other direction. The structural adjustment policies of the IMF and World Bank frequently assumed that only market mechanisms were needed to promote economic growth and that government's role had to be reduced drastically. They frequently ignored the fact that economic and institutional reforms would have to be enacted and implemented by governments, and that government had an important and legitimate role in creating the conditions that supported economic growth in the west and in newly industrializing countries of Asia. Experience showed that political and administrative institutions had to be established or strengthened before economic reform was feasible. Successful implementation of economic and political restructuring programs required far more than simply declaring new policies and removing all restrictions on private enterprise. Government officials had to be committed to the philosophy of decentralization, privatization and wider participation in decision-making, before extensive reforms could be implemented effectively and sustained over a long period of time (Waterbury 1989). Widespread political support had to be generated among national political leaders for transferring planning, decision-making and managerial authority to lower levels of administration and to the private sector (Rondinelli and Minis 1990).

Structural adjustment policies often ignored the fact that in a decentralized political system and a market-oriented economy both local governments and private enterprises need a network of institutions to support them in carrying out development activities. Provisions must be made for extending physical infrastructure, transport, and communications linkages, and for distributing the public services and facilities that make local communities efficient economic units.

Experience suggests that many of the difficulties with conventional approaches to planning and administration can only be overcome by reorienting administrative practices and procedures and the perceptions of the development administrators who work in international organizations and governments in developing countries. The challenge is to find more appropriate ways of dealing with the inevitable uncertainty and complexity of development problems and of creating the institutional diversity that allows experimentation, innovation, and widespread participation in economic activities.

The changes that are needed include:

(1) adjusting planning procedures and methods of administration to the processes of political interaction through which policies are actually enacted and carried out;
(2) adopting a learning approach to planning and administration in order to cope with uncertainty and complexity;
(3) building widespread and appropriate forms of administrative capacity among government agencies, non-governmental organizations, and the private sector within developing societies;
(4) decentralizing to the appropriate level of government and to the private sector the authority required for planning and administering development activities effectively;
(5) simplifying analysis and management procedures in international assistance organizations and governments in developing countries;
(6) encouraging rather than suppressing or punishing error detection and correction in development administration; and
(7) creating greater flexibility for development administrators in governments and international assistance organizations to cope with complexity and uncertainty by offering incentives for innovation, risk-taking and learning.

Together, these principles form an alternative approach to development management: a process of *adaptive administration.* The process is as crucial to the development of poor countries today as it has been to the development of the industrialized countries for more than two centuries. Of course, it is foolish to assume that economic growth and social development will proceed in the poor countries of Asia, Africa, and Latin America in the same way as they did in western countries. Yet, the underlying conditions of autonomy, experimentation, and diversity are as crucial for developing societies to discover the most appropriate means of achieving their development objectives as they were for western countries in finding their paths to development.

ADJUSTING PLANNING AND ADMINISTRATIVE PROCEDURES TO THE POLITICAL DYNAMICS OF PUBLIC POLICY-MAKING

No system of development administration can be effective that ignores or discounts the political dimensions of decision-making. Ultimately, all development plans are political statements and all attempts to implement them are political acts. The pretension that planners and administrators are politically objective or neutral is naive. The belief that development administration is beyond the scope of politics usually reduces planners and administrators to a politically ineffective advisory role in which they produce plans that are never implemented. The illusion that they are "above politics" sometimes leads planners and administrators in international organizations and bureaucracies in developing countries to rely predominantly or exclusively on control-oriented management practices that are at best irrelevant or at worst perverse. The politics of development planning and implementation has been given relatively little attention in development theory, and this may explain, in part, its failure to deal explicitly with the political dimensions of development assistance.

International aid organizations often assume that, because many developing countries do not have democratic or participative political systems, decisions are made by a ruling elite without conflict or political interaction, and are carried out dutifully by bureaucrats in the lower levels of the administrative hierarchy. But experience, both with development projects during the 1960s and 1970s, and with structural adjustment loans during the 1980s and early 1990s, indicates that a great deal of political conflict and interaction attend decision-making and implementation, even in authoritarian political systems (Nelson 1989). An internal evaluation of the World Bank found that, as a result of the lack of attention to political complexities, structural adjustment loans were "overly optimistic about governments' commitment to reform and their implementation capacity, and the Bank tended to envisage unmanageably broad and rapid programs of reform" (Nicholas 1988).

Experience shows that comprehensive planning and control-oriented management were rarely effective in situations where decisions were made and implemented through the interaction of groups with different interests, objectives, sources of power, and capacities to undermine or block the action of others (Rondinelli 1987). Although in Mexico elite groups and national bureaucracies dominate the political process, Grindle's (1977) studies showed that political factions, patron–client relationships, ethnic ties, and personal networks were often more influential in implementing projects and in making demands on the bureaucracy for resources and services. In Colombia these groups usually act without complete or even correct information and on motives that are often hidden or distorted by the need to mobilize a broad base of support (Bromley 1981). Rather than seeking to make and follow deliberate, well-researched, and comprehensive plans for development projects, participants allow their own actions to be shaped by the interaction of competing interest groups, by unfolding events, by external forces, and by the unpredictable reactions of private and government organizations during the course of political conflict.

Comprehensiveness, consistency, a long-range view, and cost-effectiveness – the criteria by which planners supposedly forge their proposals for development projects – are heavily discounted by political leaders under pressure to act quickly by interest groups with short-term horizons and the need to provide visible results quickly.

When development activities are funded by international agencies, the organizational environment of decision-making becomes even more complex and uncertain, and political interaction becomes more intense. The advice offered to governments in developing countries by the staff or consultants of international agencies can change the balance of political power among interest groups, expand or constrain the access of various groups to centers of decision-making, alter the government's development priorities, or generate social changes with undesired political impacts. Because of their political potency, projects and programs are often subjected to close political scrutiny by one or more international agencies, a variety of national ministries, private organizations, beneficiary groups, and powerful individuals within and outside of government.

Even organizations such as the World Bank, which are mandated by their charters to control carefully the conditions under which projects are selected and carried out, have great difficulty achieving their goals without political negotiation and compromise. In its evaluation of structural adjustment loans, the World Bank discovered that "difficulties and delays in reorganizations have been caused by the complexity of necessary procedural and legal changes, but also by vested interests and well-established bureaucracies" (Nicholas 1988). These political interests frequently obstructed and delayed the simplification of export regulations and procedures and the implementation of changes in agricultural production and distribution policies.

In a complex decision-making system, in which political commitment and support from powerful groups is a necessary but not sufficient condition for the success of development projects, systematic planning, and control-oriented administration are usually ineffective. An important challenge to development administrators, therefore, is to identify and test procedures and techniques of planning and management that rely less on control and more on incentives. In order to be more effective they must reorient their role from one of attempting to dominate and control development projects to one of helping intended beneficiaries to formulate, implement, and evaluate their own plans. In developing countries the resources of many groups with different perspectives and bits of knowledge are needed to ensure the success of development projects and economic reforms.

A variety of methods of intervention and interaction are available to influence the outcome of policy conflicts and to facilitate development activities. (see Fig. 6.1). But because development administration has traditionally been defined as technical, politically neutral, and objectively rational, alternative methods of planning and analysis are rarely explored in the literature. The methods most often used by development planners and administrators – central control and coordination – are the least effective instruments of influencing decisions and behavior.

A wide variety of methods of influence, including information dissemination, public education, specialized training programs, persuasion, and consultation techniques that require less direct intervention in policy conflicts can be used by various groups in society to affect the outcome of political decisions. In some societies, governments, political parties, and interest groups use more sophisticated methods of psychological manipulation, modeling, and techniques that psychologists call "shaping and reinforcement" to influence behavior through non-coercive means. Incentives and rewards are often used more effectively to achieve desired behavior than punishments and threats, which are the basis of all control-oriented methods of management (Rondinelli 1975, 1987).

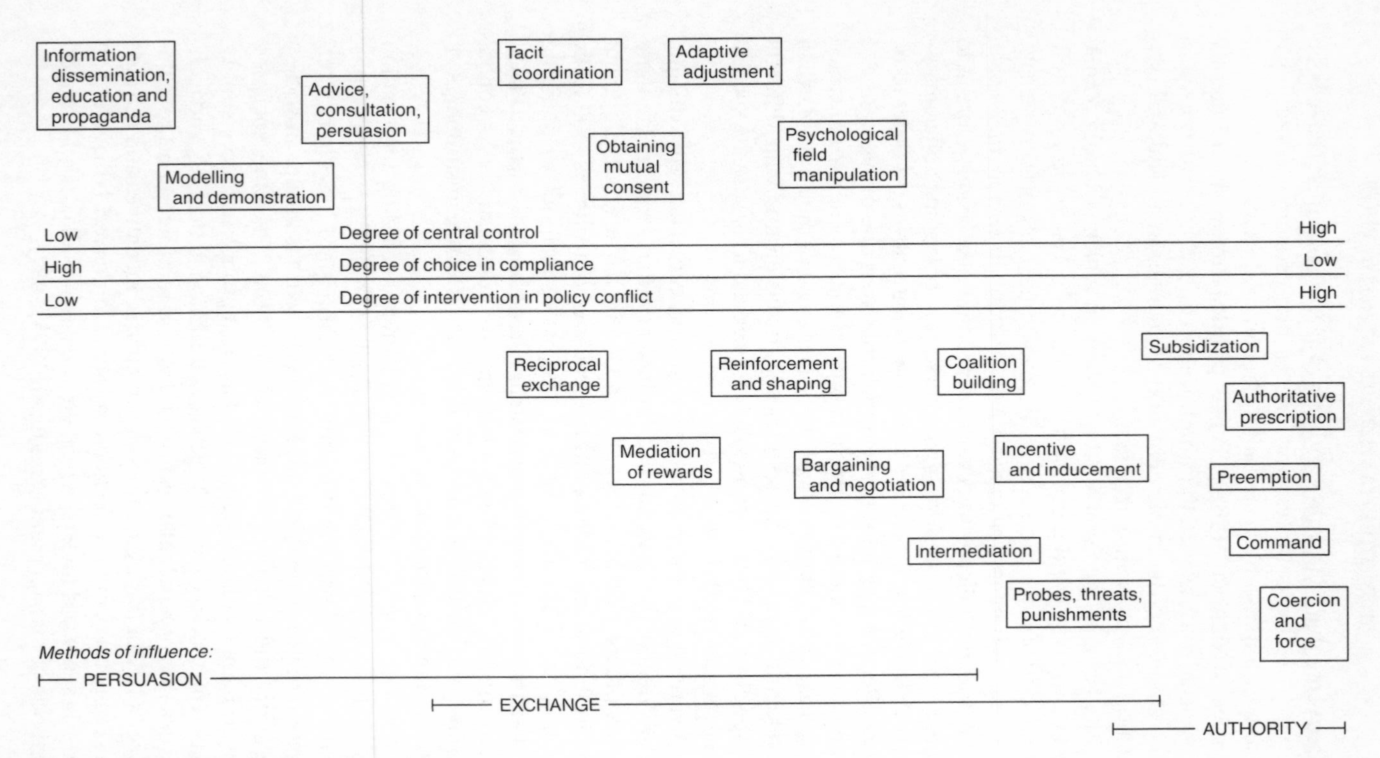

Figure 6.1 Processes of social interaction

Development administrators must give much more attention to processes of reciprocal exchange, compromise, formal and informal bargaining and negotiation, mediation, and coalition building in the process of decision-making if they are to become more effective in coping with the complexity and uncertainty of development problems. Once these means of influence are recognized, the costs of comprehensive planning, rationalistic analysis, and control-oriented management become difficult to defend.

INCREASING THE RESPONSIVENESS OF BUREAUCRACIES ENGAGED IN DEVELOPMENT ACTIVITIES

To be effective in the future, international assistance organizations and governments in developing countries must be restructured to promote the innovation and creativity needed to make them more responsive to the needs of their clientele. In unstable and uncertain environments organizations must be responsive to survive.

One means of improving the responsiveness of bureaucratic organizations is to restructure them to be more sensitive to changes in their environments and to provide incentives that reward staff who encourage participation of their clientele. Under conditions of uncertainty, organic rather than mechanistic structures are needed (Burns and Stalker 1961). An organic structure requires:

(1) breaking organizations down into task-oriented sub-units rather than specialized functional sub-units;
(2) adjusting and redefining tasks through the interaction of organizational members rather than allowing them to be rigidly defined and adjusted only by the organization's top leaders;
(3) encouraging individuals to accept broader responsibilities and commitments than those prescribed by their functional job descriptions;
(4) creating a collegial system of decision-making based on interaction among peers rather than strictly by hierarchical authority and control;
(5) recognizing the limited knowledge of formal leaders about the organization's activities and soliciting information from the organization's members and clients;
(6) creating lateral communications and consultation arrangements among people in different ranks rather than relying on vertical communications between superiors and subordinates;

(7) creating an environment in which the commitment of staff is to performing tasks and fulfilling responsibilities effectively and responsively rather than to blind loyalty and obedience to superiors;
(8) providing rewards for staff to improve their performance and develop expertise in the larger environment rather than only within the organization itself; and
(9) increasing the participation of staff and clients in decisions about changes in the organization's missions, goals, and functions.

In participatory and responsive organizations the distinction between political and administrative behavior loses much of its meaning: "managers are seen as active contributors to policy making as are all members and clients of the organization" (Thomas and Brinkerhoff 1978). Indeed, clients would participate in decision-making and management along with the staff and leadership of the organization. In an organization that seeks to be responsive, the dominant goal is to facilitate self-determination among its clients or within the community with which it is dealing. The role of development administrators is to provide support for initiatives and to make available resources that will allow beneficiaries to achieve their development objectives.

Often a less costly and more expedient alternative to organizational reform is to change the external environment of a bureaucracy to make it more competitive and to reduce the organization's dominant or monopoly power over its clients. Lamb (1982) has suggested a number of ways by which this can be done. One is by encouraging *direct competition*: that is, allowing or encouraging more public institutions or private and voluntary organizations to perform development functions or to offer goods and services previously provided by only one or a few public agencies. Often competition can be encouraged through licensing, deregulation, or subsidies. Agencies in some developing countries use traditional institutions to provide services in rural areas. Others subcontract to private firms as a means of supplementing or enhancing their own capacities. Governments often share responsibilities with private firms or allow public enterprises to provide services in order to introduce competition in the public sector (Rondinelli and Kasarda 1991).

From his experience with World Bank poverty alleviation projects, Salmen (1989: 18) has concluded that "the operation of the learning process as a method of selecting organizations for survival and growth works better if there are a plurality of organizations competing with one another." He noted that when there is only a single organization performing an important function, "one will feel constrained to keep it in existence, even when it is doing poorly."

Another market surrogate for organizational reform is *actively to market government services*: that is, to make consumer preferences more influential in decisions about service provision and delivery. Lamb has argued that government agencies should be required to:

> provide and market a convenient "shelf" of information and services, in contrast to a fixed or predetermined "package" from which users could select their desired mix. This might apply to productive services (e.g., agricultural inputs and extension), to social services (e.g., literacy, health, family planning) or to a combination of the two plus other public and private services.
> (Lamb 1982: 4)

Such an approach would require an appropriate structure for service delivery, with the kinds of characteristics that will be described later in this chapter.

A third market surrogate involves *organizing recipients, users and suppliers* so that external constituencies can bring pressure on public agencies to be more effective and responsive in performing their tasks. Unofficial or voluntary organizations can also provide goods and services on a low-cost, self-help basis, thereby lessening clients" dependence on a single public agency. The organization of recipients, users and suppliers creates a system of countervailing power that provides channels of interaction between client and agency.

Finally, *performance agreements*, in which functions are contracted or delegated to non-government agencies or to the beneficiaries themselves, are another means of introducing market surrogates in the public sector. Private enterprises or voluntary organizations would be given contractual responsibilities for performing functions, providing services, or producing goods, for as long as they perform their responsibilities effectively and responsively. Changes could be made in the agreements to ensure continued effectiveness and responsiveness without initiating massive bureaucratic reorganizations that could easily be opposed by entrenched civil servants (Rondinelli 1991).

Eventually, making the environment of public agencies more competitive would require them to reorganize internally along the lines suggested earlier. Creating competition would provide clients with alternative sources of obtaining goods and services, and give them bargaining power that would allow them to obtain more responsive and effective treatment by public agencies.

ADOPTING A LEARNING APPROACH TO PLANNING AND ADMINISTRATION

The application of adaptive administration requires planners and administrators to view social problem-solving as an incremental process of social interaction, trial and error, successive approximation, and learning. Such an approach requires an institutional culture quite different from that of Weberian bureaucracies. The recognition of these needs led in the 1970s to the institution-building approach for development administration. The theory of institution-building was based on the premise that the poor record of implementing development policies was the result of the inability of governments in developing countries to perform essential development functions. Siffin (1977: 59) argued that "the essence of development is not to maintain, but to create effectively. . . . Doing this means, among other things, marshalling substantial amounts of knowledge about organizational design and the effects of alternative organizational arrangements." Thompson (1974) insisted that hierarchical models of bureaucracy were inappropriate for performing the creative tasks required of development administration, and that development administrators needed institutions that provided an atmosphere conducive to innovation. Both tool-oriented and political reform theories sought to strengthen the central government's control over development. But because policies change rapidly in developing countries, Thompson argued that they were not susceptible to central direction. He called for institutions in which administrators could solve problems more creatively: non-hierarchical, non-bureaucratic, professional, problem-oriented systems in which communications structures were loose and in which decisions evolved from group interaction.

Esman (1969: 13) defined institution building as "the planning, structuring, and guidance of new or reconstituted organizations which embody changes in values, functions, physical and social technologies; establish, foster and protect new normative relationships and action patterns; and, obtain support and complementarity in the environment." The aim of institution building was to create viable development institutions, that is, those with the following characteristics:

(1) technical capacity – the ability to deliver technical services that are innovative for the society into which they are introduced;
(2) normative commitment – the ability to internalize innovative ideas, relationships and practices within the staff of the organization;
(3) innovative thrust – the ability of the organization to continue to innovate so that new technologies and behavior patterns will not be "frozen" in their original form;
(4) environmental image – the ability of the organization to attain favorable recognition within society and be highly valued or regarded by other organizations; and
(5) spread effects – the ability of the organization to get other institutions to adapt the innovative technologies, norms, or methods that it has introduced.

Institution-building strategy was concerned not only with strengthening the administrative capacity of individual organizations, but also with forging cooperative relationships among them.

Although the institution-building approach recognized the need for an innovative, learning-based style of administration, application was generally limited to central government ministries and to large educational and research institutes. The abstract nature of the theory made it difficult to apply in the Third World, and where it was applied the application was often considered to be an end in itself. It did not, therefore, address questions of equity and participation or seek to increase the access of the poorest groups to institutional resources.

Adaptive administration requires not only more flexible and responsive institutions, but also a learning-based process of planning and administration. Korten has argued persuasively that development planning and administration must be based more on social learning than on scientific management. It must depend less on finding rational and optimal solutions than on discovering useful actions that can ameliorate adverse conditions of individuals and groups who are affected by them and in ways that are acceptable to them. He argues that:

> the key to social learning is not analytical method, but organizational process; and the central methodological concern is not with the isolation of variables or the control of bureaucratic deviations from centrally-designed blueprints, but with effectively engaging the necessary participation of system members in contributing to the collective knowledge of the system.
>
> (Korten 1981: 613)

Korten contends correctly that more complex problems need more localized solutions and more broadly based participation in decision making. Korten and Carner (1982: 10–12) show that even large bureaucracies, such as the National Irrigation Administration (NIA) in the Philippines, can be organized in ways that allow them to be responsive to the needs of local communities and that allow participation by a wide variety of interest groups and beneficiaries in planning and managing development projects. The communal irrigation committees (CICs) that the NIA used to guide pilot irrigation projects were able to mobilize institutional resources, pull together professionals from various agencies involved in irrigation programs and local leaders into a coalition for change, and provide direction and guidance for centrally initiated but locally managed development. The CICs did not impose a set of organizational relationships on a community but made existing relationships and processes legitimate. Membership on the committee was voluntary, and members joined and left as projects progressed and as their interests dictated. The functional benefits of the committee compelled some members to remain and provide continuity. The processes of planning and management were based on experience with irrigation projects in different communities rather than controlled by centrally determined rules and regulations.

But learning-based administration implies the need to go beyond new organizational arrangements to educate and train individuals in the process of joining action with learning. John Friedmann (1973) noted that there are four essential elements of individual learning on which the process must be based. These are the ability to:

(1) question existing reality and to raise questions about existing practices, relationships, and conditions, not only to understand and cope with them more effectively but also to appreciate how and when to make needed changes;
(2) draw general lessons from particular experience, which in turn requires shared observation and inductive reasoning;
(3) test theory in practice so that actions can be infused with and guided by experience; and
(4) examine results in an objective and sincere way to uncover and examine mistakes as well as to apply successful approaches in new situations.

None of these elements, of course, is easy to apply in conventional bureaucracies or international assistance organizations, which are organized to standardize, routinize, and limit individual discretion. Finding ways of inculcating the spirit of learning, experimentation and creativity in hierarchical bureaucracies remains a challenge for development administrators.

DEVELOPING WIDESPREAD AND APPROPRIATE FORMS OF ADMINISTRATIVE CAPACITY

International assistance organizations and governments of developing countries have rarely given adequate attention to alternative ways of organizing for development. International agencies have either left governments to their own devices or insisted on administrative arrangements that conform to western administrative principles.

Experience suggests, however, that the ability of planners and managers to implement projects and programs as policy experiments depends on building administrative capacity at all levels in developing societies, both by involving a wider range of organizations and by decentralizing authority and responsibility. Esman and Montgomery (1980) have pointed out that human resource development programs require a variety of organizational arrangements for eliciting the participation of those who are to be affected by the projects, and for assuring that resources reach the intended groups. In addition to using centrally financed and managed bureaucracies, programs should be administered through modified bureaucracies that can be released from conventional procedures and central controls to extend their reach in unconventional ways.

Local governments, for example, often have linkages with local groups that central bureaucracies do not know about and cannot reach. Market mechanisms may be used more effectively to deliver services with only modest government intervention. Many voluntary organizations have ties and channels of distribution to groups that are frequently identified as beneficiaries of social and economic development projects, and through which experimental, pilot, and demonstration projects can be carried out. Finally, it is necessary to build the administrative capacity of, and involve in planning and administration, what Esman and Montgomery call "organized special publics": interest groups such as credit unions, women's clubs, irrigation associations, labor unions, and cooperatives.

By actively involving these groups in project planning and management, international organizations and governments in developing countries can create constituencies that will support projects for their members and act as channels of interaction between beneficiaries and the agencies sponsoring development programs. Once mobilized, these organizations can also begin to generate their own projects, thereby supplementing and extending the impact of those sponsored by government.

At the same time, if implementation is to be improved, organizations in developing countries must begin to practice what Jon Moris (1977: 127) calls "engaged planning." Moris agrees that development policies and programs often cannot be implemented successfully by existing bureaucracies through routine administration. Special arrangements must often be made to protect and promote new programs until they are institutionalized. He contends that in East Africa, for example, "development does not occur under either private or socialist auspices unless someone regularly puts in a large margin of extra intelligence effort of a managerial nature." Either individuals with a high degree of motivation to achieve program goals must be placed in charge of these programs or special organizations must be created to administer the programs outside of the regular bureaucratic structure. In any case, Moris argues that:

> somebody must keep the daily activities of distinct but vertically interlocked services under surveillance, must frame contingency plans . . . must indulge in bureaucratic politics in order to secure the commitments implied in action programs, and must be prepared even to break the rules in an emergency.
>
> (Moris 1977: 127)

In developing countries, trade-offs must be made between the costs of apparent redundancy and duplication and the increased probabilities of succeeding in attaining development goals. Some theorists argue that the creation of redundancy, far from being inefficient and wasteful, is essential for increasing the reliability of service delivery. Caiden and Wildavsky (1974) point out that the ability of rich countries to obtain the resources needed for production and service delivery owes less to management efficiency than to complex redundancy. When a large number of organizations with resources and skills are performing the same or similar functions, the failure of one organization is not critical. Others fill the gap, thereby greatly increasing reliability and reducing uncertainty. "Arrangements in poor countries," they note (p. 63), "lack the benefits of redundancy – the surplus, the reserve, the overlapping networks of skills and data – to cushion the reverberating effects of uncertainty."

Some governments and international assistance agencies, such as the World Bank, have attempted to avoid the constraints and weaknesses of public administration by creating autonomous implementation units for each large-scale project. These institutions were usually outside of the formal government structure and were given the resources necessary to carry out their tasks. Often they could pay higher salaries, better benefits, and amenities, have greater flexibility to innovate or act expeditiously and were able to attract the most competent staff. But the creation of large numbers of autonomous development authorities also had some disadvantages. Too often they became powers unto themselves and over time amassed enough political influence to pursue their own interests, which were often in conflict with national development policies. Some became so dependent on external financial and technical support that they responded more readily to international assistance agency priorities and to technical standards than to the needs of their own clientele.

International organizations have only begun to discover, however, that creation of separate project implementation units is no substitute for strengthening administrative capacity more widely among public and private institutions in developing countries if they are to sustain the benefits of development projects after international funding ends.

DECENTRALIZING AUTHORITY FOR DEVELOPMENT PLANNING AND ADMINISTRATION

If administrative capacity for planning and implementing development projects is to be strengthened in developing countries, then both political leaders and international assistance agencies must give much more attention to decentralizing authority, responsibility, and resources. This implies that governments must decentralize by strengthening the field units of national ministries, creating or strengthening local administrative units, delegating functions to regional, special purpose, or functional authorities, devolving responsibilities and resources to local governments, and involving the private sector in service delivery. To the extent that economic growth depends on innovation and change, it also depends on the freedom to experiment. Such freedom requires the decentralization of authority to diverse organizations in society.

Administrative decentralization can achieve a variety of goals in developing countries (Rondinelli 1990). First, it can make the implementation of national policy more effective by delegating to local officials greater responsibility for tailoring development projects to local conditions and needs. Decentralization can also make it easier for local officials, interest groups, and businesses to cut through the enormous amounts of red tape and the highly bureaucratic procedures characteristic of planning and administration in developing nations – that result in part from the over-concentration of power, authority, and resources in the central government. By decentralizing development functions to the field offices of government agencies, or to local governments, more government officials can become knowledgeable and sensitive to local problems and needs because they will be working at the level where these are most visible and pressing. Closer contact between local populations and government officials can also allow the latter to obtain better information with which to formulate plans and programs than they can obtain in the national capital.

Decentralization can also increase political and administrative support for national development policies at the local level, where government plans are often unknown by the local population or are undermined by local elites, and where support for government policies is usually weak. Decentralization can promote national unity by giving groups in different sections of the country the ability to participate in planning and decision-making and thus to increase their "stake" in maintaining political stability. Decentralization is also a way of increasing the efficiency of central agencies by relieving top management of routine, detailed tasks that could more effectively be performed by field staff or local political leaders or non-government organizations (Rondinelli 1981a, 1982).

An enduring problem of decentralization in developing countries, however, is the "paradox of power" (Rondinelli 1979a). Although strong central political commitment is necessary to initiate administrative reforms, they cannot be effectively implemented and sustained without diffuse political support and widespread participation. Those whose political commitment is necessary to initiate the reforms, however, often consider such a diffusion of power and participation as a serious political threat. Development projects that reallocate economic resources, increase income, and expand participation in the economy also create new and potentially more powerful interest groups that can make claims on and challenge central authority. Indeed, the creation of countervailing power is often a precondition for sustaining organizational reforms such as administrative decentralization and privatization.

Although some development projects need guidance and support, their implementation must be tailored to unique or special local needs, especially in areas where resources and institutional capability are lacking or weak. But serious obstacles to decentralization exist in most developing countries. Few governments have been willing to establish state, provincial, regional, or district governments with sufficient autonomy and resources to carry out their tasks or with sufficient power to coordinate the work of different ministries operating within their jurisdictions. In nearly all countries, the tensions and conflicts over control and distribution of political power are an underlying factor. The tensions between the desire to maintain central control and the need to diffuse support and elicit participation are evident in all societies attempting to decentralize administrative authority (Cheema and Rondinelli 1983, Rondinelli 1990). China's ambivalence toward decentralization both during and after Mao's regime, and the recurring shifts between delegation of authority and recentralization, clearly illustrated the "paradox of power." Ambivalence at the center resulted from fear of the growing power of provincial leaders during periods of decentralization and from their pursuit of regional self-sufficiency in economic development – sometimes in conflict with national plans and policies. Only in the early 1990s did local governments in China develop enough power to slow the process of recentralization (Chen 1990).

Without decentralization and privatization, however, it is difficult to develop the widespread administrative capacity needed to plan and manage development activities in responsive and adaptive ways. Yet the paradox of power continues to obstruct the decentralization and spread of administrative capacity that is necessary for effective experimentation and innovation both in international assistance organizations and in governments of developing countries. Reviewing the history of economic growth in western countries, Rosenberg and Birdzell correctly observe that a:

> grant of that kind of freedom costs a society's rulers their feeling of control, as if they were conceding to others the power to

determine the society's future. The great majority of societies, past and present, have not allowed it. Nor have they escaped from poverty.

(Rosenberg and Birdzell 1986: 34)

BUILDING AN EFFECTIVE INSTITUTIONAL NETWORK FOR SERVICE DELIVERY

Development depends not only on strengthening the administrative capacity of diverse organizations within society, but also on creating an effective institutional network through which public services and privately produced goods can be delivered effectively.

If international assistance organizations and governments in developing countries are to respond more effectively to the needs of those groups who traditionally have been bypassed by economic progress, three tasks must be successfully carried out: first, the hold of clientelist politics must be broken; second, it must be replaced with organizations or coalitions strong enough to represent the interests of the poor when decisions about them are made; and third, an effective institutional structure for delivering services and technology and other resources to those who need them must be established.

Experience suggests that little can be done to deliver essential services to rural areas where people are dependent on wealthy landowners or elites for their livelihood and survival, and where such dependence manifests itself in the patrons" domination of local and national politics through personalism, favoritism, and the manipulation and control of large voting blocs (Powell 1970, Scott 1972). In places where rural people depend on patrons to purchase their labor and products, provide shelter, offer credit, loans, and aid in emergencies or crises, and assist in paying for marriages and funerals, those government programs that threaten to weaken or destroy the patron–client relationship will receive no sympathy from the elite and little support from the rural population. Where dependent patron–client relationships provide whatever small amount of security rural people can expect in an uncertain and precarious life, development projects will have little impact unless they are accompanied by political change and unless the government or the market can displace the functions performed by patrons by providing demonstrably better, more responsive, more reliable services with little risk for the local population.

The difference between more effective and less effective development projects has often been the willingness and ability of government to assist in creating an organizational framework for mobilizing leadership, sharing power and decision-making, and expanding economic participation (Uphoff 1986).

In Taiwan and Japan, for example, the success of agrarian reform projects that provided the base for widespread development depended first on breaking the control of wealthy landowners and then on substituting strong local government and a network of farmers' associations, irrigation associations, land reform committees, and cooperatives that allowed participation and provided a channel of communication and influence.

The capacity of public and private organizations to deliver services to the poor was one of the most crucial factors affecting the success of rural development projects during the 1970s and 1980s. To a large extent, subsistence farmers and landless laborers were poor because they lacked access to public and private institutions that had the resources they needed to increase their productivity and incomes (Rondinelli and Ruddle 1977). In many rural regions essential services did not exist or were provided only in traditional forms. The public institutions through which social services were usually delivered were either missing or did not serve the majority of people; indeed, they often severely exploited the poor.

Moreover, institutions in rural areas were rarely linked into a network of supporting institutions to allow continuous, reliable, and efficient flows of services, or they had low levels of administrative capacity, and were unable to deal with the complex problems of development. New institutions introduced by government or international agencies were frequently so incompatible with traditional practices, customs, and behavior that they not only failed to serve, but in some cases further alienated rural people.

Although the concept of appropriate technology is well established, relatively little attention has been given to its organizational dimensions or, more importantly, to the characteristics of adaptive institutions for delivering social services and technological improvements in developing countries.

Adaptive institutions have a number of important characteristics (Rondinelli and Ruddle 1977). First, like appropriate technology, they must be responsive to diverse problems and conditions found in developing countries. Beneficiaries are often quite heterogeneous; the urban poor have different needs than rural farmers, shifting cultivators and landless laborers. Moreover, their settlement patterns are usually so dissimilar that a delivery system designed to meet the needs of only one group will overlook or inadequately serve the others. Organizational solutions can no more be designed and universally prescribed for all areas within a developing country than they can for all developing countries.

Second, adaptive institutions must be complementary and able to integrate the services of many other organizations with those that they provide. Services and technologies must be mutually reinforcing and interlocking in order to stimulate local development. Adaptive institutions must be linked both vertically and horizontally to provide a network of services and to increase the quality and reliability of delivery (Uphoff and Esman 1974).

Third, adaptive organizations must be built on culturally accepted arrangements, practices, and behavior. Understanding traditional institutions that have served people in developing countries for decades or centuries – their strengths, inadequacies, limitations, and potentials for transformation – is essential for modifying old institutional arrangements and introducing new ones. The success of agrarian reform in Taiwan and Japan was largely attributed to the modification of traditional farmers' associations for new functions in land redistribution and community development. Even after the revolution in China, Mao's planners were careful to adopt acceptable local arrangements as the basis of communal structure. The communal system, as Stavis (1974: 54–5) points out, "did not emerge from an historical and social vacuum; it was not simply proclaimed according to a vision of society." On the contrary, the communes were "intimately related to centuries of economic and political history and to almost a decade of gradually expanding organs of economic cooperation, slowly increasing in size, complexity, responsibility and degree of socialization." In much the same way, the agricultural and industrial reforms that came after Mao's death were not designed by the central government but evolved out of local practices. Leasing land for household farming, marketing of goods that exceeded the communal quotas, and the emergence of private small-scale enterprises were approved, formally or informally, by the government only after these practices became widespread (Chen *et al.* 1990).

Fourth, although they must be culturally acceptable, the institutional network must be designed to transform traditional practices and behavior into more suitable arrangements for economic growth and equitable income distribution. They must also be catalysts for change, transforming developmentally inadequate practices and behavior at a locally acceptable pace. Moreover, they must gradually displace those traditional institutions that are incapable of change and remain flexible enough themselves to be adapted to changing conditions as development occurs.

Finally, adaptive institutions must be designed in conjunction with beneficiaries and open to local participation and leadership. Efficient and effective service delivery can rarely be attained through standardized locational criteria or by mechanisms designed by professional technicians and administrators. It depends on an intimate understanding of people's varied behavior and motivations, something that is unlikely to occur without the participation of intended beneficiaries.

During the late 1980s and early 1990s international assistance organizations and governments in developing countries began to consider more seriously expanding the role of informal–sector and private enterprises in providing many services that previously could be offered only by state-owned enterprises or government bureaucracies (Rondinelli and Kasarda 1991). Advocates of privatizing public services argued that private enterprises have numerous advantages over government agencies – such as lower production costs, greater efficiency in service delivery, and greater capacity to obtain and maintain capital equipment. Under favorable conditions, the private sector can offer consumers greater choice and provide services more flexibly than government agencies. Non-governmental organizations can usually make decisions more quickly and efficiently than public bureaucracies. When they deliver services effectively, private organizations can reduce the size of the public payroll and some of the government's financial burdens while generating needed tax revenues for other public purposes (Savas 1982, Hatry 1983, Pirie 1985).

Market pressures can stimulate private firms to find ways to cut their costs and increase their competitiveness and productivity. Private firms are usually less restricted in work and hiring practices than public agencies and can use labor more productively. Businesses can, when necessary, reduce their work force more easily than governments can, and are able to lower labor costs per unit of output. Even contracting-out has advantages: for example, it allows governments to take advantage of specialized skills in non-governmental organizations. By contracting competitively for services, governments can determine the true costs of production and thereby eliminate waste. Contracting also permits governments to adjust the size of programs incrementally as demands or needs change.

World Bank studies point out that:

> in Brazil, Kenya and Argentina, urban road and highway maintenance can be done more effectively under contract and at considerable cost savings. For example, in Ponta Grossa, Brazil, road maintenance was 59 percent more costly when done by municipal workers than when done by a private contractor.
>
> (World Bank 1986: 30)

Contracting with small-scale enterprises using donkey carts to collect garbage in the Central Region of Eastern Sudan was also found to be a more efficient and less costly alternative to purchasing expensive imported garbage trucks that could only collect periodically from scattered town dustbins. This public–private partnership resulted in costs that were only 10 percent of truck service. Moreover, nearly 26 percent of these costs could be recovered through household charges (Lewis and Miller 1986).

Studies of privatization in other countries show not only cost reductions but also efficiency gains. World Bank (1986) studies point out that in many Third World cities small private bus companies and minibus operators provide better – and often less costly – transport services than do large municipal bus systems. Moreover, public bus systems often do not offer convenient or flexible service in slum and squatter settlements. Nor do they run major transport routes within easy walking distance of low-income neighborhoods. In Istanbul, Calcutta, Bangkok, and Nairobi, private minibuses provide more convenient and flexible service in and around poor neighborhoods. In some cities, low-income families prefer private transportation even when fares are somewhat higher than those of public transport.

Increasingly, developing countries must seek the most appropriate and effective means of delivering public services that support productive activities and look to the private sector as a crucial part of the institutional network for service delivery.

RELYING ON ADJUNCTIVE AND STRATEGIC RATHER THAN COMPREHENSIVE AND CONTROL-ORIENTED PLANNING

In their (now classic) study of planning and budgeting procedures in developing nations, Caiden and Wildavsky (1974: 293) concluded that "if we were asked to design a mechanism for decisions to maximize every known disability and minimize any possible advantage of poor countries, we could hardly do better than comprehensive, multi-sectoral planning." Drawing on reviews of planning and budgeting processes in 80 developing countries they confirmed many of the assertions made in Chapters 2 and 3 of this study, that comprehensive planning "calls for unavailable information, non-existent knowledge, and a political stability in consistent pursuit of aims undreamed of in their existence. Thus, this kind of planning turns the most characteristic features of poor countries into obstacles to development."

Improving the administration of complex and uncertain development activities requires new forms of planning. Adjunctive and strategic planning must be substituted for comprehensive analysis and central control. Adjunctive planning seeks to facilitate decision-making among a wide variety of organizations and interests in society, focus attention on solving remediable aspects of known problems, identify courses of action that move marginally, incrementally and through successive approximation away from unsatisfactory social and economic conditions even when "optimal" or ideal goals cannot be agreed upon, and explore alternatives on which diverse interests can act jointly (Lindblom 1965, Rondinelli 1971, 1975).

Lindblom (1965) has described three underlying principles of strategic analysis and planning. First, analysis is limited to courses of action that produce incremental changes in existing conditions, moving away from undesirable states toward those on which varying political interests can agree. This makes the tasks of analysis and planning more manageable in complex, uncertain, or risky situations. Strategic planning begins with what is known and attempts to broaden the base of knowledge and to formulate alternative interventions that will set other changes in motion, rather than beginning with sweeping and comprehensive reforms, the effectiveness and feasibility of which cannot be predicted.

Second, strategic planning seeks to convert large, complex development problems into smaller, disaggregated, ones that can be dealt with incrementally. This reduces the complexity of analysis, allowing better use of information and permitting a reconsideration of goals and means as unanticipated forces appear during implementation. Breaking up large development problems into components that can be dealt with through smaller-scale projects has strong advantages. Chambers (1978: 211) has pointed out that big projects can be a trap. Once initial commitments are made to undertake large, complex projects, forces are often set in motion that make the commitments irreversible and that overwhelm planners and administrators after the projects are underway. Moreover, decisions are more easily constrained and choices foreclosed in large projects than in smaller ones. Chambers notes that "the 'yes–no' decision begins to close and often closes before any formal cost–benefit analysis can be carried out."

Caiden and Wildavsky had earlier (1974) concluded from their analysis of planning and budgeting in poor countries that large projects "encase probable errors in concrete." They argued that breaking down development activities into smaller projects could facilitate strategic planning and increases the possibilities for learning, adaptation and correction. They contended that:

> Because less has been invested in each individual project, it is relatively less expensive to end them. Since all are not supposed to survive, there is less concern that each will have to be continued or prove productive. Damage can be contained in size and scope. By investing future resources in similar ventures, the good can be built upon. The multiplicity of efforts, the redundancy if you will, increases likelihood that a variety of approaches to the same problem will be tried and that some will work. Projects thus are used, not to carry out existing knowledge, likely to be lacking, but to obtain knowledge through action.
>
> (Caiden and Wildavsky 1974: 309)

Smaller or less complex projects require less information, less coordination, and less demand for highly skilled or experienced managers. Because they are less costly, they can be allowed to fail and "may be looked at as hypotheses that can be falsified by unfolding events, thus improving the chance for future projects to stand the test of experience."

Studies of USAID and UNDP small-enterprise development assistance programs also suggest (Liedholm and Mead 1986, Kilby 1979) that projects providing assistance to small-scale enterprises were most effective when they provided only a single "missing ingredient" to firms that could otherwise operate effectively. Programs that were task-specific and tailored to the needs of particular industries or product groups were more successful than those attempting to help large numbers of disparate enterprises. Those that were formulated on the basis of surveys of the industries to be assisted in order to discover the "missing ingredients" needed, and the effective demand for them, tended to be more successful than general purpose programs or those with poorly defined clientele.

Third, strategic and adjunctive planning focus on the examination of goals in close connection with emerging values and with the availability of resources to achieve them, and they encourage interaction with the groups that must participate in or benefit from projects. This requires planners to examine alternatives in light of the values embedded in current behavior and attitudes, cultural tradition, and economic and political constraints rather than formulating strategies in the "splendid isolation" of the headquarters of international assistance organizations or national planning agencies.

SIMPLIFYING PLANNING AND MANAGEMENT PROCEDURES

It is also clear from past experience that administrative procedures and arrangements for development must be relatively simple and uncomplicated. Complex managerial methods rarely work at any level of government in developing countries. People either ignore complex administrative procedures or are exploited by government officials who can manipulate them. Moreover, skills and resources for management are in short supply in most poor countries and administrative capacity will remain relatively weak. Chambers and Belshaw concluded from their experience with the management of rural development programs in Kenya that:

> in designing management procedures, the temptation is to introduce more and more requirements and measures, more and more complicated techniques and more and more elaborate relationships. But such an approach quickly leads to a drop in output and eventually to paralysis.
>
> (Chambers and Belshaw 1973: 6.8)

As noted in Chapters 4 and 5, the tendency of international assistance organizations to insist on sophisticated and complicated methods of analysis, planning, and programming often creates more problems than it solves. Amin (1978: 230) has noted that there is a tendency for economists to equate complexity with sophistication. But, "true sophistication might lie in seeing ways in which one could simplify, or short-circuit a long route by adopting a much shorter and simpler plan." Often there is a need for more common sense rather than more rigorous analyses. Defining common sense as the "store of information which men collect through their various experiences but which is very difficult to spell out, let alone quantify," Amin (1978: 232) maintains that it is captured in the judgment or intuition of experienced administrators and leaders.

Less complicated techniques of planning and analysis can provide greater opportunities for both experienced administrators and beneficiary groups to bring their knowledge to bear on project selection and implementation. Chambers (1978) suggests displacing sophisticated techniques of cost–benefit appraisal with simple decision matrices, ranking systems that focus on the effects of alternative projects on intended beneficiaries, checklists of desired project characteristics, and approximate cost comparisons rather than elaborate rate-of-return and cost–benefit studies. In view of the high degree of uncertainty and the paucity and unreliability of data in most developing countries, such simplified methods of analysis may not only be quite adequate but may be more optimal and cost-effective than complex and rationalistic methods. Analytical techniques might then be perceived as aids to judgment rather than as substitutes for it.

ENCOURAGING ERROR DETECTION AND CORRECTION RATHER THAN SUPPRESSION AND PUNISHMENT

One of the most difficult changes to make in the bureaucracies of both international development agencies and developing country governments will be in administrators" attitudes toward error. Under conditions of uncertainty, errors and mistakes are not only likely, they are to be expected. The concept of development policy-making as social experimentation requires that projects be designed in such a way that errors and mistakes can be uncovered as a project proceeds; it can then be redesigned and revised incrementally. Yet the bureaucratic systems in most developing countries and international organizations are designed to suppress mistakes and errors and to punish managers when mistakes are discovered.

Such attitudes are not only unrealistic but dangerous. They cause administrators and managers to be fearful of making mistakes, encourage them to cover up or deny the existence of errors, and discourage correction, redesign, and redirection. They inhibit creativity, innovation, flexibility, and experimentation – the very stuff of successful project management. Moreover, they prevent planners and administrators from involving other groups and organizations in decision-making, thereby obstructing the process of gathering the information and tapping the knowledge needed to make projects more relevant and responsive. Because of their fear of error detection, managers avoid evaluation and monitoring, instead of using these procedures as tools of adaptive administration.

Learning by doing becomes more acceptable only when administrators explicitly recognize that all development projects are basically social experiments. Incentives and rewards must be provided to encourage innovation, experimentation, and creativity. Only uncorrected errors should be considered evidence of poor management. The ability of managers and administrators to monitor and evaluate continuously, so that mistakes are uncovered and corrected quickly, must be strengthened in the bureaucracies of developing countries and international assistance organizations.

CREATING INCENTIVES FOR INNOVATIVE MANAGEMENT

Administrative systems and procedures are needed that encourage, support, and reward those managers who take responsibility for development activities – especially in poor countries where uncertainty and risk of failure are greater and where the knowledge required to plan and implement projects effectively is least likely to be available. The unwillingness of bureaucrats in most developing countries to work outside of the national capital and in local and provincial governments is due not only to a civil service system that is structured to reward service in central ministries and agencies, but also because civil servants perceive that their specialized skills would not be appreciated or rewarded in the provinces.

Attempts to decentralize development planning and administration in many developing countries over the past two decades, for example, were strongly opposed by technicians and professionals in central ministries who feared that in the provinces they would not get adequate support from local officials (Rondinelli 1982, 1990).

The reluctance of central government officials to serve in the provinces is also attributable to the differences in living conditions between the national capital and smaller cities and towns, where housing is often considered inadequate, educational opportunities for children are limited, supplies and equipment are difficult to obtain, and there are few opportunities for professional advancement. The inability or unwillingness of national governments to provide suitable living conditions and incentives for civil servants to work in the provinces threatens the success of decentralized development projects and severely weakens the capacity of the central governments to provide needed technical and managerial support to local organizations.

Cleaves (1980) concluded, after reviewing a variety of urban and rural development programs in Zambia, Peru, India, Colombia, Mexico, Kenya and Brazil, that another essential element in improving administration in developing countries is that "national leaders and policy makers must change their frame of reference as to the *definition of personal and policy success* and reward policy implementors accordingly." He argued that administrators who are assigned to risky projects or remote areas must be specially and visibly rewarded when they carry out their tasks successfully:

> When money is limited, quality personnel in outlying districts scarce, and possibilities for change marginal, administrators who undertake these assignments must be recompensed, in terms of remuneration and prestige, for handling small budgets, working in difficult terrains, and accomplishing small, gradual and continuous change. The policy must contain self-evident measures of social improvement so that administrators can record their progress. While appeals for altruism are legitimate ways to build motivation, they cannot completely substitute for direct compensation, especially when implementors sense that they are bearing the brunt of responsibility for national development.
>
> (Cleaves 1980: 296)

An essential part of the incentive system must be adequate training, a strong system of field support and greater recognition of the special conditions under which field staff work. Esman and Montgomery (1980) have found that the field staff who work with the poor and those populations living in remote areas are usually the worst trained, the least supervised, the weakest motivated, and rarely supported by efficient supply systems. The capacity of field administrators to work with beneficiary groups must be strengthened and their ability to mobilize human and financial resources in ways that foster mutual respect and cooperation between government administrators and local residents must be improved. In many countries, this means finding ways of changing the attitudes and behavior of government officials toward the public and toward the purposes of their jobs. The work of administrators should be facilitating and supporting rather than control oriented. Field administrators must be trained to recognize the capacity of local residents, regardless of their social status, level of education, or income, to make important contributions of knowledge, skills, or commitment to the success of development activities.

Moreover, after field administrators and project managers are trained, they must be given discretion to manage without the rigid supervision that smothers flexibility and creativity. If they are to be effective they must be given adequate resources to obtain the equipment, supplies, personnel and facilities needed to carry out their tasks. Upward mobility within the civil service system must be tied to field performance. As long as field service posts are perceived of as inferior, unrewarded, and dead-end positions, ambitious and skilled managers will resist field assignments and continue to seek positions at headquarters in the national capital.

CONCLUSIONS

Sustainable and equitable economic development requires strengthening administrative capacity throughout developing societies. It implies expanding participation in economic activities, strengthening the capacity of a wide variety of public and private organizations to plan and carry out development activities, and increasing the access of individuals to resources and opportunities needed to meet their basic human needs, raise their productivity, and develop their potential. Courses of action that lead to the attainment of these objectives will remain complex and uncertain. Planning and managing development activities seeking to attain these objectives require an adaptive approach.

Arguments that adaptive administration cannot be used effectively in large bureaucracies is becoming less convincing as more studies of projects that were planned and managed in a participatory and collaborative manner with local organizations become available (Uphoff 1986). Adaptive administration and participatory planning have been used successfully for agricultural and rural development projects in Central America and Asia (Korten and Alfonso 1981). Flexible and adaptive planning was used for water supply programs in Malawi, irrigation projects in Sri Lanka and the Philippines, and rural development projects in Bangladesh, Thailand, India, and other countries (Hafner and Rosenweig 1984). The success of community-managed water-supply projects depended on local groups taking responsibility for ownership and maintenance of water systems and the government granting local communities authority to make decisions and assume control over the operations (McCommon *et al.* 1990). The Grameen Bank in Bangladesh succeeded by allowing the beneficiaries themselves to make decisions and organize self-help activities. In Latin America, a large number of health, education, community service, and small business projects implemented by local organizations and private voluntary groups using participatory approaches have succeeded where large-scale government or internationally funded projects have had questionable results (Gran 1983, Ashe 1985).

Increasing evidence indicates that these methods can also be used by large organizations. The United Nations Children's Fund (UNICEF) (1982) has been using adaptive administration and participatory planning in its urban basic services projects in Sri Lanka, India, Peru, Indonesia, Mexico, Malaysia, Ethiopia, Ecuador, and several Central American countries, often with strong support from their governments. These projects were identified, planned, and formulated collaboratively by community groups, government officials, and UNICEF advisors. Services were provided at low cost through self-help. Most project staff were selected by the community in which they worked, and their activities were tailored to the conditions and needs of beneficiaries. The programs were planned and implemented concurrently and redesigned frequently as more was learned about their impacts.

Also, the Inter-American Foundation (IAF) has made small grants for nearly twenty years to local private groups that help the poor improve their social and economic conditions. The IAF supports a wide variety of self-help experimental programs and projects, bypassing central governments and working directly with the poor. The beneficiaries themselves take the primary responsibility for project identification and planning and for management and control of the projects" implementation. The IAF keeps its administrative costs low and works with a minimum of red tape; it approves or rejects project proposals quickly and follows up with supervision and technical assistance when it is requested in a low-key but effective way (Bell 1984).

Dealing effectively and responsively with development problems on a larger scale in the future will depend on the ability of planners and administrators to use more effectively the political, social, and economic mechanisms of authority, exchange, and persuasion, which Lindblom (1977) argues are the fundamental means of influencing behavior. Development planners and administrators must not only understand these mechanisms better but also learn how to combine them in new ways to cope with complexity and uncertainty. Lindblom's observation of India applies as well to many other developing countries:

> India's difficulties in economic development are in part the consequences of her leaders" inability to understand that growth requires a growth mechanism; if not the market, which Indian policy cripples, then the authority of government, which India has never chosen to mobilize.
>
> (Lindblom 1977: 6)

In either case, effective development administration is unlikely to emerge from conventional principles, which emphasize comprehensive, detailed and control-oriented planning and management. In an uncertain and complex world, planning must be participatory and administration must be adaptive. These approaches require managers who can facilitate rather than control the interaction of those individuals and groups who have the bits of knowledge, resources, and experience needed to change undesirable conditions and help define more acceptable courses of action.

Adaptive administration requires skilled people who can act as catalysts, mobilizing those whose support or commitment is needed to make development programs and projects relevant, appropriate, and successful. It demands administrators who can respond creatively to the needs of intended beneficiaries and quickly to changes in conditions that affect the success of development activities. It requires people who are willing and able to seek out and correct mistakes as they are discovered, and who can plan and manage simultaneously. It requires administrators who view themselves as leaders rather than as bureaucrats. It calls for managerial systems in international organizations and governments of developing countries that train administrators to join action with learning, to experiment, to test new ways of doing things and to be sensitive and responsive to the needs of the people they serve. Most of all, it requires administrative systems that let good managers manage, and that reward them for their efforts and results.

Finding new ways of establishing conditions that allow development administrators to recognize and cope effectively with the inevitable complexity and uncertainty of development problems will be the strongest challenge for developing countries and international assistance agencies in the future.

REFERENCES

Adamolekun, L. (1988) "Political leadership in Sub-Saharan Africa: from giants to dwarfs", *International Political Science Review* 9, 2: 95–106.

Adelman, I. (1975) "Growth, income distribution and equity-oriented development strategies", *World Development* 2–3, 3: 67–76.

—— (1986) "A poverty focused approach to development policy", in J.P. Lewis and V. Kallab (eds) *Development Strategies Reconsidered*, Washington: Overseas Development Council, 49–65.

—— and Morris, C. (1967) *Society, Politics and Economic Development*, Baltimore: Johns Hopkins University Press.

—— (1973) *Economic Growth and Social Equity in Developing Countries*, Stanford: Stanford University Press.

Ahmad, Y. (1977) "Project identification, analysis and preparation in developing countries: a discursive commentary", in D. Rondinelli (ed.) *Planning Development Projects*, Stroudsburg, Pa.: Hutchinson & Ross, 161–5.

Amin, G. (1978) "Themes from the discussion at the symposium", *World Development* 6, 2: 227–40.

APHA (1977) *The State of the Art of Developing Low Cost Health Services in Developing Countries*, Washington: American Public Health Association.

Ashe, J. (1985) *The PISCES II Experience, Vol I: Local Efforts in Micro-Enterprise Development*, Washington: US Agency for International Development.

Ayres, R. (1984) *Banking on the Poor: The World Bank and World Poverty*, Cambridge, Mass.: MIT Press.

Bacha, E. (1987) "IMF conditionality: conceptual problems and policy alternatives", *World Development* 15, 12: 1469–82.

Bainbridge, J. and Sapire, S. (1974) *Health Project Management*, Geneva: World Health Organization.

Baum, W. (1978) "The World Bank project cycle", *Finance and Development* 15, 4: 10–17.

Bell, P. (1984) "Testimony of the Inter-American Fund president", US Congress, House of Representatives, Foreign Assistance and Related Program Appropriations for 1984, 98th Congress, 1st Session, Washington: Government Printing Office: 81–111.

Benveniste, G. (1972) *The Politics of Expertise*, Berkeley, California: Glendessary Press.

Berg, E. (1989) "The liberalization of rice marketing in Madagascar", *World Development* 17, 5: 719–28.

Braybrooke, D. and Lindblom, C. (1970) *A Strategy of Decision*, New York: Free Press.

Bromley, R. (1981) "From cavalry to white elephant: a Colombian case of urban renewal and marketing reform", *Development and Change* 12: 77–120.

—— and Bustelo, E. (1982) "Introduction", in Bromley and Bustelo (eds), *Politica y Technica No Planajamento: Perspectivas Criticas*, Brasilia: United Nations Children's Fund, 9–11.

Brown, D. (1971) *Agricultural Development in India's Districts*, Cambridge, Mass.: Harvard University Press.

Brown, R. (1986) "International responses to Sudan's economic crisis: 1978 to the April 1985 coup d'etat", *Development and Change*, 17, 3: 487–511.

Burki, S. (1980) "Sectoral priorities for meeting basic needs", *Finance and Development* 17, 1: 18–22.

Burns, T. and Stalker, G. (1961) *The Management of Innovation*, London: Tavistock.

Caiden, N. and Wildavsky, A. (1974) *Planning and Budgeting in Poor Countries*, New York: Wiley.

Callaghy, T. (1989) "Toward state capability and embedded liberalism in the Third World: lessons for adjustment", in J.M. Nelson (ed.) *Fragile Coali- tions: The Politics of Economic Adjustment*, New Brunswick, N.J.: Trans- action Books, 115–38.

Cernea, M. (1989) "User groups as producers in participatory afforestation strategies", Development Discussion Paper no. 319, Cambridge, Mass.: Harvard Institute for International Development.

—— (1991) "Using knowledge from social science in development projects", World Bank Discussion Paper no. 114, Washington: World Bank.

Ceylon, Government of (1959) *The Ten Year Plan for Ceylon*, Colombo: Government Press.

Chambers, R. (1978) "Project selection for poverty focused rural development: simple is optimal", *World Development*, 6: 209–19.

—— and Belshaw, D. (1973) *Managing Rural Development: Lessons and Methods from East Africa*, Brighton: Institute of Development Studies, University of Sussex.

Chand, G. (1989) "The World Bank in Fiji: The case of the Suva-Nadi highway reconstruction project", *Development and Change* 20, 2: 235–68.

Changrien, P. (1970) *Development Planning in Thailand*, unpublished Ph.D. dissertation, Ann Arbor, Michigan: University Microfilms.

Cheema, G.S. and Rondinelli, D. (1983) "Introduction" in Cheema and Rondinelli (eds), *Decentralization and Development: Policy Implementation in Developing Countries*, Beverly Hills, Ca.: Sage Publications.

Chen, K. (1990) "The failure of recentralization in China", Research Paper Series no. 6, Country Economics Department, Washington: World Bank.

—— Jefferson, G. and Singh, I. (1990) "Lessons from China's economic reform", Research Paper Series no. 5, Country Economics Department, Washington: World Bank.

Chenery, H. (1974) "Introduction", in H. Chenery, M. Ahluwalia, C. Bell, J. Duloy and R. Jolley (eds) *Redistribution with Growth*, London: Oxford University Press, xiii–xx.

Choldin, H. (1969) "The development project as a natural experiment: the Comilla, Pakistan project", *Economic Development and Cultural Change* 17: 483–500.

Cleaves, P. (1980) "Implementation amidst scarcity and apathy: political power and policy design", in M. Grindle (ed.), *Politics and Policy Implementation in The Third World*, Princeton: Princeton University Press, 281–303.

Cohen, J. (1974) "Rural change in Ethiopia: the Chilalo agricultural development unit", *Economic Development and Cultural Change* 22: 580–614.

—— (1979) *Integrating Services for Rural Development*, Discussion Paper, Cambridge, Mass: Harvard Institute for International Development.

—— (1987) *Integrated Rural Development: The Ethiopian Experience and the Debate*, Uppsala, Sweden: The Scandinavian Institute of African Studies.

Cointreau, S. (1982) *Environmental Management of Urban Solid Wastes in Developing Countries*, Washington: World Bank.

Cole, D. and Lyman, P. (1971) *Korean Development: The Interplay of Politics and Economics*, Cambridge, Mass.: Harvard University Press.

Conyers, D., Warren, D.M. and Van Tilberg, P. (1988) "The Role of integrated rural development projects in developing local institutional capacity", Ames, Iowa: Iowa State University Research Foundation.

Corbo, V. and de Melo J. (1987) "Lessons from the Southern Cone policy reforms", *The World Bank Research Observer* 2, 2: 111–42.

Cowen, R. and McLean, M. (1984) *International Handbook of Educational Systems*, vol. 3, Chichester, England: Wiley.

Crane, B. and Finkel, J. (1981) "Organization impediments to development assistance: The World Bank's population program", *World Politics* 33, 4: 516–33.

Cuca, R. and Pierce, C. (1977) *Experiments in Family Planning: Lessons from the Developing World*, Baltimore: Johns Hopkins University Press.

Currie, L. (1966) *Accelerating Development*, New York, McGraw-Hill.

Das, N. (1973) "India's planning experience", *World Today*, October: 430–9.

Davis, J. (1974) *An Introduction to Public Administration*, New York: Free Press.

de la Barra, X. and Moretti, M. (1988) "CEO study of UNDP-supported technical cooperation in the field of housing", New York: United Nations Development Program.

Demery, L. and Addison, T. (1987) "Stabilization policy and income distribution in developing countries", *World Development* 15, 12: 1483–98.

Eaves. P. (1989) "OED analysis of institutional development", Washington: World Bank.

Esman, M. (1969) "Institution building as a guide to action", unpublished paper, Washington: US Agency for International Development.
—— (1972) *Administration of Development in Malaysia*, Ithaca: Cornell University Press.
—— (1977) "Monitoring the progress of projects: the redbook and operations room in Malaysia", in D. Rondinelli (ed.) *Planning Development Projects*, Stroudsburg, Pa.: Hutchinson & Ross, 225–9.
—— and Montgomery, J. (1980) "The administration of human development", in P. Knight (ed.) *Implementing Programs of Human Development*, World Bank Staff Working Paper no. 403, Washington: World Bank, 183–234.
Ewing, A. (1974) "Pre-Investment", *Journal of World Trade Law* 8: 316–28.
Ferris, J. and Graddy, E. (1986) "Contracting out: For what? With whom?" *Public Administration Review* 46, 4: 332–44.
Fields, G. (1988) "Employment and economic growth in Costa Rica", *World Development* 16, 12: 1493–509.
Frankel, F. (1971) *India's Green Revolution*, Princeton, N.J.: Princeton University Press.
Friedman, M. (1958) "Foreign economic aid: means and objectives", in G. Ranis (ed.) *The United States and the Development Economies*, New York, Norton, 250–63.
Friedmann, J. (1973) *Retracking America: A Theory of Transactive Planning*, Garden City, New York: Anchor Press.
Gittinger, J. (1972) *Economic Analysis of Agricultural Projects*, Baltimore: Johns Hopkins University Press.
Gordenker, L. (1976) *International Aid and National Decisions: Development Programs in Malawi, Tanzania and Zambia*, Princeton: Princeton University Press.
Gran, G. (1983) "Learning from development success: Some lessons from contemporary case histories", NASPAA Working Paper No. 9, Washington: National Association of Schools of Public Affairs and Administration.
Grant, J. (1973) "Development: the end of trickle down?", *Foreign Policy* 12: 42–65.
Griffin, K. and Enos, J. (1970) "Foreign assistance: objectives and consequences", *Economic Development and Cultural Change* 18: 313–27.
—— and Khan, A. (1978) "Poverty in the Third World: ugly facts and fancy models", *World Development* 6, 3: 295–304.
Grindle, M. (1977) *Bureaucrats, Politicians and Peasants in Mexico*, Berkeley: University of California Press.
Gurley, J. (1974) "Rural development in China", in E. Edwards (ed.) *Employ- ment in Developing Nations*, New York: Columbia University Press, 383–403.
Hafner, C. and Rosensweig, F. (1984) "Water and sanitation for health projects: case study, Malawi", paper presented at Workshop on Development Management, New York: Annual Conference of the American Society for Public Administration.
Hapgood, D. (1965) *Policies for Promoting Agricultural Development*, Cambridge, Mass.: MIT Center for International Studies.
Haque, W., Mehta, N., Rahman, A. and Wignaraja, P. (1977) "Micro level

development: design and evaluation of rural development projects", *Development Dialogue* 2: 71–137.

Harbison, F. (1962) "Human resources development planning in Modernizing Economies", *International Labour Review* 5: 2–23.

Hardoy, J. (1990) "Ex-post evaluation of UNDP Projects on urbanization: case studies in Argentina", New York: United Nations Development Program.

Hatry, H. (1983) *A Review of Private Approaches for Delivery of Public Services*, Washington: The Urban Institute.

Heginbotham, S. (1975) *Cultures in Conflict: The Four Faces of Indian Bureauracracy*, New York: Columbia University Press.

Helleiner, G. (1987) "Stabilization, adjustment and the poor", *World Development* 15, 2: 1499–513.

Henderson, P. (1980) "What's wrong with the Brandt Report?: Economics askew", *Encounter* December: 12–17.

Hirschman, A. (1959) *The Strategy of Economic Development*, New Haven, Conn: Yale University Press.

—— (1967) *Development Projects Observed*, Washington: The Brookings Institution.

Hogan, M. (1987) *The Marshall Plan: America, Britain and the Reconstruction of Western Europe 1947–1952*, Cambridge, England: Cambridge University Press.

Honadle, G., Morss, E., Van Sant, J. and Gow, D. (1980) "Integrated rural development: making it work?", Washington: Development Alternatives Incorporated.

Hoos, I. (1972) *Systems Analysis in Public Policy: A Critique*, Berkeley, Ca: University of California Press.

Hoselitz, B. (1964) "Advanced and underdeveloped countries: A study in development contrasts", in W. Hamilton (ed.) *The Transfer of Institutions*, Durham, N.C.: Duke University Press, 27–58.

Humphrey, D. (1962) "Indonesia's national plan for economic development", *Asian Survey* 2, 10: 12–21.

Hussain, I. (1973) "Mechanics of development planning in Pakistan: a suggested framework", *Pakistan Economic and Social Review* 2, 4: 454–62.

Hutchinson, E., Montrie, C., Hawes, J. and Mann, F. (1974) *Intercountry Evaluation of Agricultural Sector Programs*, vol. 4, Washington: US Agency for International Development.

Ickis, J. (1981) "Structural responses to new rural development strategies", in D. Korten and F. Alfonso (eds) *Bureaucracy and the Poor: Closing the Gap*, Singapore: McGraw-Hill, 4–32.

Independent Commission on International Development (Brandt Report) (1980) *North–South: A Program for Survival*, Cambridge, Mass.: MIT Press.

Institute of Development Studies (IDS) (1972) *An Overall Evaluation of the Special Rural Development Program*, Occasional Paper no. 8, Nairobi: University of Nairobi.

International Labor Organization (ILO) (1976) *Tripartite World Conference on Employment, Income Distribution and Social Progress and the Inter-*

national Division of Labor: Declaration of Principles and Program of Action, Geneva: ILO.
—— with ARTEP (1987) *Structural Adjustment – By Whom, For Whom?* New Delhi, India: ILO, Asian Employment Program.
Jackson, R. (1969) *A Study of the Capacity of the United Nations System*, Geneva: United Nations.
Johnston, B. and Clark, W. (1982) *Redesigning Rural Development: A Strategic Perspective*, Baltimore: The Johns Hopkins University Press.
—— and Mellor, J. (1961) "The role of agriculture in economic development", *American Economic Review* 51, 4: 571–81.
Kamarck, A. (1983) *Economics and the Real World*, Philadelphia, Pa.: University of Pennsylvania Press.
Karunatilake, H. (1971) *Economic Development in Ceylon*, New York: Praeger.
Katz, S. (1970) "Exploring a systems approach to development administration", in F. Riggs (ed.) *Frontiers of Development Administration*, Durham, N.C.: Duke University Press, 109–38.
Kean, J., Turner, A., Wood, D. and Wood, J. (1987) *Analysis of Institutional Sustainability Issues in USAID: 1985–86 Project Evaluation Reports*, Washington: US Agency for International Development.
Khan, A. (1978) "Pakistan — the Daudzai experiment", in Asian Productivity Organization, *Rural Development Strategies of Selected Member Countries*, Tokyo: APO, 139–45.
Khan, M. (1987) "Macroeconomic adjustment in developing countries: a policy perspective", *The World Bank Research Observer* 2, 1: 23–42.
Kilby, P. (1979) "Evaluating Technical Assistance", *World Development* 7: 309–23.
Kim, J.K. (1978) "Republic of Korea", in Asian Productivity Organization, *Rural Development Strategies of Selected Member Countries*, Tokyo: APO, 12–20.
Kim, K. and Kim, O. (1977) "Korea's Saemaul Undong: social structure and the role of government in integrated rural development", *Bulletin of Population and Development Studies Center* 6: 1–15.
Korea, Republic of (1971) *The Third Five Year Economic Development Plan 1972–1976*, Seoul: Government Printer.
—— (1980) *Saemaul Undong 1980*, Seoul: Ministry of Home Affairs.
Korten, D. (1980) "Community organization and rural development: a learning process approach", *Public Administration Review* 40, 5: 480–511.
—— (1981) "Management of social transformation at national and subnational levels", unpublished paper, Manila: Ford Foundation.
—— and Alfonso, F. (eds) (1981) *Bureaucracy and the Poor: Closing the Gap*, Singapore: McGraw Hill.
—— and Carner, G. (1982) "Reorienting bureaucracies to serve people: two experiences from the Philippines", unpublished paper, Manila: Ford Foundation.
Krueger, A. (1987) *The Importance of Economic Policy in Development: Contrasts Between Korea and Turkey*, Cambridge, Mass: National Bureau of Economic Research.

—— (1989) "The role of multilateral lending institutions in the Development Process", *Asian Development Review* 7, 1: 1–20.

Kulaba, S. (1990) "Thematic evaluation of UNDP urban projects in Tanzania", New York: United Nations Development Program.

Kulp, E. (1977) *Designing and Managing Basic Agricultural Programs*, Bloomington, Indiana: Indiana University International Development Institute.

Kuznets, S. (1966) *Modern Economic Growth*, New Haven, Conn.: Yale University Press.

Lal, D. (1987) "The political economy of liberalization", *The World Bank Economic Review* 1, 2: 273–99.

Lamb, G. (1982) "Market-surrogate approaches to institutional development", unpublished paper, Washington: World Bank.

Landau, M. (1970) "Development administration and decision theory", in E. Weidner (ed.) *Development Administration in Asia*, Durham, N.C.: Duke University Press, 73–103.

Langois, R. (ed.) (1986) *Economics as a Process: Essays in the New Institutional Economics*, Cambridge: Cambridge University Press.

LaNier, R., Massoni, A., and Oman, C. (1986) "Public and private partnerships in housing: a background paper", Washington: US Agency for International Development.

Lankatilleke, L. (1989) "Colombo: the integration of urban livelihoods in the Million Houses Programme Settlements", *Regional Development Dialogue* 10, 4: 141–58.

LaPorte, R., Jr. (1970) "Administrative, political and social constraints on economic development in Ceylon", *International Review of Administrative Sciences* 35, 158–71.

Ledesma, A. (1976) *Land Reform Programs in East and Southeast Asia: A Comparative Approach*, Research Paper no. 69, Madison, Wis.: University of Wisconsin Land Tenure Center.

Lee, E. (1977) "Development and income distribution: a case study of Sri Lanka and Malaysia", *World Development* 5, 4: 279–89.

Lee, H.B. (1970) "The role of the higher civil service under rapid social and economic change", in E. Weidner (ed.) *Development Administration in Asia*, Durham, N.C.: Duke University Press, 107–31.

Lee, H.I. (1969) "Project selection and evaluation", in I. Adelman (ed.) *Practical Approaches to Development Planning*, Baltimore: Johns Hopkins University Press, 241–52.

Lehman, H. (1990) "The politics of adjustment in Kenya and Zimbabwe", *Studies in Comparative International Development* 25, 3: 37–72.

Lele, U. (1975) *The Design of Rural Development: Lessons from Africa*, Baltimore: Johns Hopkins University Press.

Lemarchands, C. and Niro, C. (1989) "Review of the habitat program financed by UNCDF in the least developed countries", New York: United Nations Capital Development Fund.

Leonard, D. (1977) *Reaching the Peasant Farmer*, Chicago: University of Chicago Press.

Levy, B. (1989) "The design and sequencing of trade and investment policy reform: an institutional analysis", Washington: World Bank.

Lewis, A. (1954) "Economic development with limited supplies of labor", reprinted in A. Agarwala and S. Singh (eds) *The Economics of Underdevelopment* New York: Oxford University Press, 1970, 440–9.
—— (1955) *The Theory of Economic Growth*, London: Allen and Unwin.
Lewis, M. and Miller, T. (1986) "Public–private partnership in African urban development", Washington: US Agency for International Development.
Liedholm, C. and Mead, D. (1986) "Small-scale industries in developing countries: empirical evidence and policy implications", East Lansing, Mich.: Michigan State University.
Lim, D. (1973) "Malaysia", in Y. Hoong (ed.) *Development Planning in Southeast Asia*, Singapore: Regional Institute for Higher Education and Development.
Lindblom, C. (1965) *The Intelligence of Democracy: Decision-Making Through Mutual Adjustment*, New York: Free Press.
—— (1975) "The sociology of planning: thought and social interaction", in M. Bornstein (ed.) *Economic Planning East and West*, Cambridge, Mass.: Ballinger, 23–60.
—— (1977) *Politics and Markets*, New York: Basic Books.
Lowry, K. (1990) "UNDP urban sector projects in Indonesia: a case study", New York: United Nations Development Program.
Maithreyan, B. (1965) "Savings in a developing economy – certain suggestions for reform and national coordination", *Economic Bulletin for Asia and the Pacific* 26, 1: 188–201.
Majone, G. and Wildavsky, A. (1978) "Implementation as evolution", in H. Freeman (ed.) *Policy Studies Review Annual*, Vol. 2, Beverly Hills, Ca: Sage Publications, 103–17.
Mangahas, M. and Subido, C. (1976) "Development planning, appraisal and performance evaluation with special reference to the Philippines", *Economic Bulletin for Asia and the Pacific* 27, 1: 13–27.
Marzouk, G. (1972) *Economic Development and Policies: Case Study of Thailand*, Rotterdam: Rotterdam University Press.
Mason, E. and Asher, R. (1973) *The World Bank Since Bretton Woods*, Washington, D.C.: The Brookings Institution.
McCommon, C., Warner, D. and Yohalem, D. (1990) *Community Management of Rural Water Supply and Sanitation Services*, WASH Technical Report No. 67, Washington: Water and Sanitation for Health Project.
McGuire, M. and Ruttan, V. (1990) "Lost Directions: US foreign assistance policy since 'New Directions' ", *Journal of Developing Countries* 24, 2: 127–80.
McNamara, R. (1973) *Address to the Board of Governors*, Washington: World Bank Group.
Meltsner, A. (1976) *Policy Analysts in the Bureaucracy*, Berkeley, Ca.: University of California Press.
Merton, R. (1940) "Bureaucratic structure and personality", *Social Forces* 18: 560–8.
Mikesell, R. (1968) *The Economics of Foreign Aid*, Chicago: Aldine.
Montgomery, J. (1972) "Allocation of authority in land reform programs: a comparative study of administrative processes and outputs", *Administrative Science Quarterly* 17: 62–75.

Morawetz, D. (1977) *Twenty Five Years of Economic Development, 1950–1975*, Washington: World Bank.
Moris, J. (1977) “The transferability of western management concepts and programs: an East African perspective”, in L. Stifel, J. Colemen and J. Black (eds) *Education and Training for Public Sector Management in Developing Countries*, New York: The Rockefeller Foundation, 73–83.
—— (1981) *Managing Induced Rural Development*, Bloomington, Indiana: Indiana University International Development Institute.
Morss, E., Hatch, J., Mickelwaite, D. and Sweet, C. (1975) *Strategies for Small Farmer Development*, Washington: Development Alternatives, Inc.
Moskowitz, K. (1982) “Korean development and Korean studies – a review article”, *Journal of Asian Studies* 42, 1: 63–90.
Mueller, P. and Zevering, K. (1969) “Employment promotion through rural development: a pilot project in Western Nigeria”, *International Labour Review* 2: 111–30.
Murphy, D., Baker, B. and Fisher, D. (1974) *Determinants of Project Success*, Chestnut Hill, Mass: Boston College Institute of Management.
Mussa, M. (1987) “Macroeconomic policy and trade liberalization: some guidelines”, *The World Bank Research Observer* 2, 1: 61–77.
Myrdal, G. (1957) *Rich Lands and Poor: The Road to World Prosperity*, New York: Harper & Row.
—— (1970) *An Approach to the Asian Drama*, New York: Vintage Books.
Nelson, J. (1984) “The politics of stabilization”, in R. Fineberg and V. Kallab (eds) *Adjustment Crisis in the Third World*, New Brunswick, N.J.: Transaction Books, 99–118.
—— (ed.) (1989) *Fragile Coalitions: The Politics of Economic Adjustment*, New Brunswick, N.J.: Transaction Books.
Nelson, M. (1973) *The Development of Tropical Lands: Policy Issues in Latin America*, Baltimore: Johns Hopkins University Press.
Nelson, R. and Winter, S. (1982) *An Evolutionary Theory of Economic Change*, Cambridge, Mass.: Belknap Press.
Nicholas, P. (1988) “The World Bank's lending for adjustment: an interim report”, World Bank Discussion Paper no. 34, Washington: World Bank.
Niehoff, R. (1977) “Some key operational generalizations and issues in the use of nonformal education”, in R. Niehoff (ed.) *Nonformal Education and the Rural Poor*, East Lansing: Michigan State University College of Education.
Nigam, S. (1975) *Employment and Income Distribution Approach in Development Plans of African Countries*, Addis Ababa: International Labor Organization.
Nophaket, S. (1973) *The Administration Requirements of Development Planning in Thailand*, unpublished Ph.D. dissertation, Ann Arbor, Michigan, University Microfilms.
Noranitipadungkarn, C. (1977) *Bangkok Metropolitan Immediate Water Improvement Program*, Honolulu: East–West Center.
Nunberg, B. (1989) “Review of public sector management issues in structural adjustment lending”, Washington: World Bank.
Nurkse, R. (1953) *Problems of Capital Formation in Underdeveloped Countries*, London: Oxford University Press.

Onibokun, A. (1990) "An evaluation of UNDP assisted projects in the urban sector in Nigeria, 1971–1985", New York: United Nations Development Program.

Onunkwo, E. (1973) *Sensitivity Analysis in Evaluation of Agricultural Projects in Nigeria*, unpublished Ph.D. dissertation, Ann Arbor, University Microfilms.

Ostrander, F. (1974) "Botswana nickel-copper: a case study in private investment's contribution to economic development", in J. Barrat (ed.) *Accelerated Development in Southern Africa*, New York: St. Martin's Press, 534–49.

Overholt, W. (1986) "The rise and fall of Ferdinand Marcos", *Asian Survey* 26, 11: 1137–63.

Paine, S. (1976) "Balanced development: Maoist conception and Chinese practice", *World Development* 4, 4: 277–304.

Paolillo, C. (1975) *A Basic Needs Strategy of Development: Staff Report on World Employment Conference*, Washington: US Government Printing Office.

Paul, S. (1988) "Institutional Aspects of Sector Adjustment Operations: Summary of Findings", Washington: World Bank.

—— (1989) "Institutional development at the sectoral level: a cross sectoral review of World Bank Projects", Washington: World Bank.

Pearson, L. (1969) *Partners in Development* (The Pearson Commission), New York: Praeger.

Pirie, M. (1985) "Privatisation: the facts and fallacies", Sydney, Australia: Centre 2000 Ltd.

Poats, R. (1972) *Technology for Developing Nations: New Directions for US Technical Assistance*, Washington: The Brookings Institution.

Powell, J. (1970) "Peasant society and clientelist politics", *American Political Science Review* 44, 2: 411–25.

Prybyla, J. (1979) "Changes in the Chinese economy: an interpretation", *Asian Survey* 19, 5: 409–35.

Pye, L. (1965) "The concept of political development", *Annals of the Academy of Political and Social Science* 358: 1–13.

Pyle, D. (1980) "From pilot project to operational program in India: the problems of transition", in M. Grindle (ed.) *Politics and Policy Implementation in the Third World*, Princeton: Princeton University Press, 123–44.

Quick, S. (1980) "The paradox of popularity: ideological program implementation in Zambia", in M. Grindle (ed.) *Politics and Policy Implementation in the Third World*, Princeton: Princeton University Press, 40–63.

Rahim, S. (1977) "Nonformal Aspects of Commilla Project", in R. Niehoff (ed.) *Nonformal Education and the Rural Poor*, East Lansing, Michigan State University College of Education, 54–68.

Ralston, L., Anderson, J. and Colson, E. (1981) "Voluntary efforts in decentralized management", Working Paper, Program on Managing Decentralization, Berkeley: University of California, Institute of International Studies.

Rana, P. (1974) "The Nepalese economy: problems and prospects", *Asian Survey* 14, 7: 651–62.

Reyes, R. and Jopillo, S. (1986) *An Evaluation of the Philippine Participatory Communal Irrigation Program*, Quezon City, Philippines: Institute of Philippine Culture, Ateneo de Manila University.

Riggs, F. (1970) "Introduction", in F. Riggs (ed.) *Frontiers of Development Administration*, Durham, N.C.: Duke University Press, 3–37.

Rondinelli, D. (1971) "Adjunctive planning and urban development policy", *Urban Affairs Quarterly* 7, 1: 13–39.

—— (1975) *Urban and Regional Development Planning: Policy and Administration*, Ithaca, N.Y.: Cornell University Press.

—— (1976a) "Public planning and political strategy", *Long Range Planning* 9, 2: 75–82.

—— (1976b) "International requirement for project preparation: aids or obstacles to development planning?", *Journal of the American Institute of Planners* 43, 3: 314–26.

—— (1977) "Planning and implementing development projects: an introduction", in D. Rondinelli (ed.) *Planning Development Projects*, Stroudsburg, Pa., Hutchinson and Ross.

—— (1978) "National investment planning and equity policy in developing countries: the challenge of decentralized administration", *Policy Sciences* 10, 1: 45–74.

—— (1979a) "Administration of integrated rural development: the politics of agrarian reform in developing countries", *World Politics* 31, 3: 389–416.

—— (1979b) "Designing International development projects for implementation", in G. Honadle and R. Klauss (eds) *International Development Administration*, New York: Praeger, 21–52.

—— (1979c) "Planning development projects: lessons from developing countries", *Long Range Planning* 12, 3: 48–56.

—— (1981a) "Government decentralization in comparative perspective: theory and practice in developing countries", *International Review of Administrative Sciences* 47, 2: 133–45.

—— (1981b) "Administrative decentralization and economic development: Sudan's experiment with devolution", *Journal of Modern African Studies* 19, 4: 595–624.

—— (1982) "The dilemma of development administration: uncertainty and complexity in control oriented bureaucracies", *World Politics* 35, 1: 43–72.

—— (1983a) "Implementing decentralization programs in Asia: a comparative analysis", *Public Administration and Development* 3, 3: 181–207.

—— (1983b) *Secondary Cities in Developing Countries: Policies for Diffusing Urbanization*, Beverly Hills, Ca.: Sage Publications.

—— (1987) *Development Administration and US Foreign Aid Policy*, Boulder, Colorado: Lynne Reinner Publishers.

—— (1990) *Decentralizing Urban Development Programs: A Framework for Analysis*, Washington: US Agency for International Development.

—— (1991) "Decentralizing water supply services in developing countries: factors affecting the success of community management", *Public Administration and Development*, 11, 4: 415–30.

—— and Kasarda, J. (1991) "Privatizing public services in developing countries: what do we know?" *Business in the Contemporary World* 3, 2: 102–13.

——, Middleton, J. and Verspoor, A. (1990) *Planning Education Reforms in Developing Countries: The Contingency Approach*, Durham, N.C.: Duke University Press.

—— and Minis, H. (1990) "Administrative restructuring for economic adjustment: decentralization policy in Senegal", *International Review of Administrative Sciences* 56, 3: 447–66.

—— and Montgomery, J. (1990) "Managing economic reform: an alternative perspective on structural adjustment policy", *Policy Sciences* 23, 1: 73–93.

—— and Ruddle, K. (1977) "Local organization for integrated rural development: implementing equity policy in developing countries", *International Review of Administrative Sciences* 63, 1: 20–30.

—— (1978) *Urbanization and Rural Development: A Spatial Policy for Equitable Growth*, New York: Praeger.

—— Middleton, J. and Verspoor, A. (1990) *Planning Education Reforms in Developing Countries: The Contingency Approach*, Durham, N.C.: Duke University Press.

Rosenberg, N. and Birdzell, L., Jr. (1986) *How the West Grew Rich: The Economic Transformation of the Industrial World*, New York: Basic Books.

Rosenstein-Rodan, P. (1943) "Problems of industrialization of Eastern and South Eastern Europe", reprinted in A. Agarwala and S. Singh (eds) *The Economics of Underdevelopment*, New York: Oxford University Press, 1970, 245–55.

Rostow, W. (1952) *The Process of Economic Growth*, New York, Norton.

Ruddle, K. and Rondinelli, D. (1983) *Transforming Natural Resources for Human Development*, Tokyo, United Nations University.

Rudner, M. (1975) *Nationalism, Planning and Economic Modernization in Malaysia*, Beverly Hills, Ca.: Sage Publications.

Ruttan, V. (1975) "Integrated rural development programs: a skeptical perspective", *International Development Review*, 18, 4: 9–16.

Salmen, L. F. (1989) "Institutional dimensions of poverty reduction", Washington, D.C.: World Bank.

Sapolski, H. (1972) *The Polaris Missile System*, Cambridge, Mass.: Harvard University Press.

Savas, E.S. (1982) *Privatizing the Public Sector*, New York: Chatham House.

Schick, A. (1973) "A death in the bureaucracy: the demise of federal PPB", *Public Administration Review* 33, 2: 146–56.

Schlesinger, J. (1968) "Systems analysis and the political process", *The Journal of Law and Economics* 11: 281–98.

Schultz, T. (1964) *Transforming Traditional Agriculture*, New Haven, Conn: Yale University Press.

Schulz, L. (1972) *Politics and Development Planning in Indonesia*, unpublished Ph.D. dissertation, Ann Arbor, Mich.: University Microfilms.

Scott, J. (1972) "Patron–client politics and political change in Southeast Asia", *The American Political Science Review* 46, 1: 91–113.

Seers, D. (1965) "The limits of the special case", *Bulletin of the Oxford Institute of Economics and Statistics* 25, 2: 77–98.

Self, P. (1975) *Econocrats and the Policy Process: The Politics and Philosophy of Cost–Benefit Analysis*, Boulder, Col.: Westview Press.

Shaefer-Kehnert, W. (1977) "Approaches to the Design of Agricultural projects", Bloomington, Indiana: Indiana University International Development Institute.

Siffin, W. (1977) "Two decades of public administration in developing countries", in L. Stifle, J. Coleman and J. Black (eds) *Education and Training for Public Sector Management in Developing Countries*, New York: Rockefeller Foundation, 49–60.

Smith, T. (1974) *East Asian Agrarian Reform: Japan, Republic of Korea, Taiwan and the Philippines*, Hartford, Conn.: John C. Lincoln Institute.

Spengler, J. (1963) "Bureaucracy and Economic Development", in J. La-Palombara (ed.) *Bureaucracy and Political Development*, Princeton: Princeton University Press, 199–232.

Stavis, B. (1974) *Peoples Communes and Rural Development in China*, Special Series on Rural Local Government, no. 2, Ithaca: Cornell University Center for International Studies.

Strachan, H. (1978) "Side effects of planning in the aid control system", *World Development* 6, 4: 467–78.

Streeten, P. (1972) *Frontiers of Development Studies*, New York: Halstead.

—— (1987) "Structural adjustment: a survey of issues and options", *World Development* 15, 12: 1469–82.

—— and Burki, S. (1978) "Basic needs: some issues", *World Development* 6, 3: 411–21.

Sussman, G. (1980) "The pilot project and the choice of an implementing strategy: community development in India", in M. Grindle (ed.) *Politics and Policy Implementation in the Third World*, Princeton: Princeton University Press, 103–22.

Taylor, K. (1970) "The pre-investment function in the international development system", *International Development Review* 12, 2: 2–10.

Tendler, J. (1976) "International evaluations of small farmer organizations: Ecuador and Honduras", Washington: US Agency for International Development.

Thimm, H. (1979) *Development Projects in the Sudan*, Tokyo: United Nations University.

Thomas, J. (1974) "Development institutions, projects and aid: a case study of the water development program in East Pakistan", *Pakistan Economic and Social Review* 12: 87–103.

Thomas, T. and Brinkerhoff, D. (1978) "Devolutionary strategies for development administration", *SICA Occasional Papers*, no. 8, Washington: American Society for Public Administration, Section on International and Comparative Administration.

Thompson, V. (1961) *Modern Organization: A General Theory*, New York: Alfred A. Knopf.

—— (1974) "Administrative objectives for development administration", *Administrative Science Quarterly* 9: 91–108.

Tokman, V. (1988) "Urban employment: research and policy in Latin America", *CEPAL Review* 34: 109–26.

Trapman, C. (1974) *Change in Administrative Structures: A Case Study of Kenyan Agricultural Development*, London, Overseas Development Institute.

Unakul, S. (1969) "Annual planning in Thailand", *Economic Bulletin for Asia and the Far East* 20, 1, 68–80.

United Nations Center for Human Settlements (UNCHS) (1987) "Lessons learned from evaluations and their usage", Nairobi: United Nations Center for Human Settlements.

United Nations Children's Fund (UNICEF) (1982) "Urban basic services: reaching children and women of the urban poor", Document no. E/ICEF/L.1440, New York: United Nations.

United Nations Development Program (UNDP) (1969) *An Evaluation of UNDP Assistance of Uganda*, New York: United Nations.

—— (1972) *UNDP Assistance Requested by the Government of Thailand for the Period 1972–1976*, New York: United Nations.

—— (1973a) *The UNDP Program in Nigeria: Report of the Evaluation Mission*, New York: United Nations.

—— (1973b) *UNDP Assistance Requested by the Government of Mexico for the Period 1973–1977*, New York, United Nations.

—— (1973C) "Investment follow-up guidelines", mimeographed, New York: United Nations.

—— (1974) *UNDP Operational and Financial Manual*, New York: United Nations.

—— (1979) *Rural Development: Issues and Approaches for Technical Co-operation*, Evaluation Study no. 2, New York, United Nations.

—— (1983) *Human Resource Development for Primary Health Care*, Evaluation Study no. 9, New York: United Nations.

—— (1988) *Development of Rural Small Industrial Enterprises: Lessons from Experience*, New York: United Nations.

United Nations Economic Commission for Africa (1969) "Development planning and economic integration of Africa", *Journal of Development Planning* 1: 109–56.

United Nations Economic Commission for Asia and the Far East (1969) "The planning and financing of social development in the ECAFE Region", *Economic Bulletin for Asia and the Far East* 20, 1: 4–37.

United States Agency for International Development (USAID) (1970) *Brazil – Education Sector loan II*, Washington: USAID.

—— (1972a) *An Evaluation of the Management of Technical Assistance: Projects in Three African Countries*, Washington: USAID.

—— (1972b) "Comments on sector analysis and sector loans in Latin America", unpublished paper, Washington: USAID.

—— (1972c) "Sector Lending in Latin America", unpublished paper, Washington: USAID.

—— (1973a) *Implementation of New Directions in Development Assistance*, Report for the US Congress, House Committee on International Relations, Washington: Government Printing Office.

—— (1973b) *Report of Operations Appraisal of East Africa*, Washington: USAID.

—— (1979a) *Country Development Strategy Statement, FY 1981: Pakistan*, Washington: USAID.

—— (1979b) *Country Development Strategy Statement, FY 1981: Sri Lanka*, Washington: USAID.

—— (1979c) *Country Development Strategy Statement, FY 1981: Burma*, Washington: USAID.

—— (1983) *Annual Budget Submission to the Congress*, Washington: USAID.

—— (1987) *Annual Budget Submission to the Congress*, Washington: USAID.

—— (1988) *Sustainability of Development Programs: A Compendium of Donor Experience*, A.I.D. Program Evaluation Discussion Paper no. 24, Washington: USAID.

—— (1989) *Development and the National Interest: US Economic Assistance into the 21st Century*, Washington: USAID.

United States Code and Congressional and Administrative News (1973) Vol. 2.

United States Congress, House of Representatives (1989) "Report of the task force on foreign assistance to the committee on foreign affairs", 101st Congress, 1st Session, Congress Document 101–32, discussion draft, mimeographed.

United States General Accounting Office (1979) *US Development Assistance to the Sahel – Progress and Problems*, Report B-159652, Washington: Government Printing Office.

Uphoff, N. (1986) *Local Institutional Development: An Analytical Sourcebook with Cases*, West Hartford, Conn.: Kumarian Press.

—— (1990) "Paraprojects as new modes of international development assistance", *World Development* 18, 10: 1401–11.

—— and Esman, M. (1974) *Local Organization for Rural Development: An Analysis of the Asian Experience*, Ithaca: Cornell University Center for International Studies

—— and Ilchman, W. (1972) "Development in the perspective of political economy", in Uphoff and Ilchman (eds) *The Political Economy of Development*, Berkeley, Ca.: University of California Press.

——, Meinzen-Dick, R. and St. Julien, N. (1986) "Improving policies and programs for farmer organization and participation in irrigation water management", Ithaca, New York: Consortium for International Development, Cornell University.

Valdepenas, V. (1973) "Philippines", in Y. Hoong (ed.) *Development Planning in Southeast Asia*, Singapore: Regional Institution for Higher Education and Development, 262–6.

Vepa, R. (1977) "Implementation: the problem of achieving results", in D. Rondinelli (ed.) *Planning Development Projects*, Stroudsburg, Pa., Hutchinson & Ross, 169–90.

Villanueva, D. (1971) "A survey of the financial system and the savings-Investment Process in Korea and the Philippines", *Finance and Development* 2: 16–19.

Wade, N. (1974) "Green revolution: a just technology, often unjust in use", *Science* 186: 1093–6; 1186–8.
Warriner, D. (1964) "Land reform and economic development", in C. Eicher and L. Witt (eds) *Agriculture in Economic Development*, New York: McGraw Hill, 280–90.
Waterbury, J. (1989) "The political management of economic adjustment and Reform", in J. Nelson (ed.) *Fragile Coalitions: The Politics of Economic Adjustment*, New Brunswick, N.J.: Transaction Books, 39–56.
Waterston, A. (1965) *Development Planning: Lessons of Experience*, Baltimore: Johns Hopkins University Press.
—— (1971) "An operational approach to development planning", *International Journal of Health Studies* 1, 3: 233–52.
Weiss, W., Waterston, A. and Wilson, J. (1977) "The design of agricultural and rural development projects", in D. Rondinelli (ed.) *Planning Development Projects*, Stroudsburg, Pa.: Hutchinson and Ross, 95–139.
Weissman, S. (1990) "Structural adjustment in Africa: insights from the experiences of Ghana and Senegal", *World Development* 18, 12: 1621–34.
Wildavsky, A. (1969) "Rescuing policy analysis from PPBS", *Public Administration Review* 29, 2: 189–202.
—— (1979) *Speaking Truth to Power: The Art and Craft of Policy Analysis*, Boston: Little Brown.
World Bank (1967) *Annual Report 1967*, Washington: World Bank.
—— (1974) *Policies and Operations*, Washington: World Bank.
—— (1975) *Rural Development Sector Policy Paper*, Washington: World Bank.
—— (1978) *Annual Review of Project Performance Audit Results*, Washington: World Bank.
—— (1979) *World Development Report 1979*, Washington: World Bank.
—— (1980) *World Development Report 1980*, Washington: World Bank.
—— (1981) *World Development Report 1981*, Washington: World Bank.
—— (1983) *Learning by Doing: World Bank Lending for Urban Development 1972–1982*, Washington: World Bank.
—— (1986) *Urban Transport: A World Bank Policy Study*, Washington: World Bank.
—— (1988) *World Development Report 1988*, Washington: World Bank.
—— (1990a) *Guide to International Business Opportunities*, Washington: World Bank.
—— (1990b) *World Development Report 1990*, Washington: World Bank.

INDEX